A Hetero-functional Graph Theory for Modeling Interdependent Smart City Infrastructure

Wester C. H. Schoonenberg • Inas S. Khayal
Amro M. Farid

A Hetero-functional Graph Theory for Modeling Interdependent Smart City Infrastructure

Springer

Wester C. H. Schoonenberg
Thayer School of Engineering
Dartmouth College
Hanover, NH, USA

Inas S. Khayal
Geisel School of Medicine
Dartmouth College
Lebanon, NH, USA

Amro M. Farid
Thayer School of Engineering
Dartmouth College
Hanover, NH, USA

ISBN 978-3-319-99300-3 ISBN 978-3-319-99301-0 (eBook)
https://doi.org/10.1007/978-3-319-99301-0

Library of Congress Control Number: 2018952912

This Springer imprint is published by the registered company Springer Nature Switzerland AG
The registered company address is: Gewerbestrasse 11, 6330 Cham, Switzerland

From Wester to my sister Vivien
From Inas & Amro to our daughters
Amina & Ayah

Preface

Cities have always played a prominent role in the prosperity of civilization. Indeed, every great civilization we can think of is associated with the prominence of one or more thriving cities. And so understanding cities—their inhabitants, their institutions, their infrastructure—what they are and how they work independently and together—is of fundamental importance to our collective growth as a human civilization. But the twenty-first century city is different. No longer do they primarily exist as urban islands in a sea of rural life but they are now ever-more connected with other cities around the world in a network of urbanized population centers. The rise of globalization as an economic activity has brought about a globalized metropolitan culture to the point where many city dwellers are more likely to hop from one city to another than to explore the wilderness of rural life. The twenty-first century is also giving rise to ***mega-cities*** with more than ten million inhabitants each. While more than 40 cities worldwide hold this distinction at present, population growth and urbanization are set to create many more. And of course, the twenty-first century is infused with ubiquitous technology that fundamentally affects how society, political and economic institutions, and infrastructure develop and function. Perhaps such ***smart cities*** are best viewed as the ultimate ***engineering system***: A class of systems characterized by a high degree of technical complexity, social intricacy, and elaborate processes aimed at fulfilling important functions in society.

Why This Book?

With all of these fundamental changes in the nature of cities in the twenty-first century, it is only reasonable to ask if our collective knowledge of engineering, science, and social science is enough to understand how this new type of complex engineering system works. To truly appreciate an answer to this question, we have to understand how we got here.

The nineteenth and twentieth centuries were marked with new and unprecedented inventions like the generator, the telephone, and the automobile. These technologies developed simultaneously along two fundamentally orthogonal trajectories. In the ***reductionist*** direction, these technologies developed as individual products with ever greater speed and precision. Many such devices grew in size to achieve economies of scale and others were miniaturized to even the nanometer length scale. Our knowledge of the ***physical sciences*** grew accordingly to meet the needs of innovation in technological marketplaces. In the ***integrationist*** direction, these individual technology products were connected to form many of the large-scale infrastructure networks we know today: the power grid, the communication infrastructure, and the transportation system. Over time, these networked systems developed even more interactions while continuing to incorporate many new technologies like solar panels, smartphones, and electric vehicles. Consequently, our knowledge of the ***informatic sciences*** developed to keep track of the tremendous information required to represent these engineering systems of ever greater scope and function.

Two such informatic sciences are particularly relevant here. ***Systems engineering***, and more recently model-based systems engineering, emerged as a practical and interdisciplinary engineering discipline that enables the successful realization of complex systems from concept, through design, to full implementation. It is well equipped to deal with systems of ever-greater complexity, be they for the greater interaction within these systems or because of the expanding heterogeneity they demonstrate in their structure and function. Notable human achievements like sending a man to the moon or landing on the surface of Mars can certainly be attributed to the effective practice of systems engineering. Despite these achievements, model-based systems engineering, however, relies till today on graphical modeling languages that provide limited quantitative insight. In contrast, ***network science*** has emerged as a scientific discipline for quantitatively analyzing networks that appear in fields across the natural, social, and engineering sciences. And yet, network science, due to its reliance on graphs as a data structure, was often unable to address the explicit heterogeneity often encountered in the systems engineering field. Even the network science developments into ***multi-layer networks*** have been recognized to have significant limitations in modeling networked systems of arbitrary topology. Despite these methodological differences, these two informatic sciences have often tackled similar intellectual challenges. For example, both fields have contributed immensely to the knowledge of ***system life-cycle properties*** like centrality, modularity, flexibility, sustainability, and resilience.

These two informatic sciences now face an even greater challenge. Not only are individual products being connected to form infrastructure networks, but these networks are forming interactions between each other to form systems-of-systems. The "smart grid," the energy-water nexus, and the electrification of transport are all good examples where one network system has fused with another to form a new and much more capable system. This trend is only set to continue. Taken to its finality, it leads us to the pressing need to understand and implement smart cities as a platform upon which to integrate all of these efforts. Naturally, the methodological and theoretical limitations of model-based systems engineering and network science must be overcome to gain truly novel insight into the development of smart cities.

It is in this context that we have written this book. *A Hetero-functional Graph Theory for Modeling Interdependent Smart City Infrastructure* lays a theoretical foundation that intellectually resembles a fusion of model-based systems engineering and network science. Hetero-functional graph theory relies on multiple graphs as data structures so as to support quantitative analysis. It also explicitly embodies the heterogeneity of conceptual and ontological constructs found in model-based systems engineering. Its application to interdependent smart city infrastructures of arbitrary topology presents a highly demanding use case.

Where Did Hetero-functional Graph Theory Come from?

While hetero-functional graph theory can be viewed as an intellectual fusion of model-based systems engineering and network science, its origins are found elsewhere. Although not called as such at the time, the theory originated from the automated mass-customized production system literature. Production systems, as their own class of engineering system, present some unique modeling challenges. They can have a nearly arbitrary size, an unlimited diversity of production capabilities, and an almost infinite number of product variants. They also demonstrate a consistently changing structure and behavior. The need to compete in dynamic marketplaces with product variants of increasingly short product life-cycle drove mass-customized production systems to explicitly foster ***reconfigurability*** as a life-cycle property of their integrated automation solutions. To that end, Prof. Farid's doctoral dissertation *reconfigurability measurement in automated manufacturing systems* (2007) specifically developed a quantitative measure of reconfigurability. Perhaps unsurprisingly to a network scientist, it used a ***design structure matrix*** as a type of graph to address the ease of reconfiguration. Furthermore, and unsurprisingly to a systems engineer, it addressed the allocation of function to form as the ***central question of engineering design***. It drew the concept of a ***knowledge base*** from the ***Axiomatic Design*** literature and quantified it as a bipartite graph. Beyond these considerations, however, the reconfigurability measurement of mass-customized production systems needed to specifically address ***heterogeneity*** as its essential characteristic. Finally, because automation was an essential aspect of mass-customized production systems, the research was explicitly ***cyber-physical***.

In 2010, the Laboratory for Intelligent Integrated Networks of Engineering Systems (LIINES) was founded with a research program devoted to the sustainability and resilience of intelligent energy systems. Several research themes were launched year after year: first smart power grids, then energy-water nexus, then electrified transportation systems, and then industrial energy management. Each of these represented engineering systems where two intelligent energy systems were integrated. As work in each research theme developed, a pattern emerged. While each new application had its peculiarities that required enhancements to hetero-functional graph theory, the graph structures originally developed in the Farid dissertation could be used generically across multiple application domains. It is around this time that the LIINES research really began to ***converge*** and several publications sought to specifically state the cross-domain applicability of hetero-functional graph theory. The article entitled *Static Resilience of Large Flexible Engineering Systems: Axiomatic Design Model and Measures* (2015) specifically demonstrated cross-domain applicability, addressed resilience as a life-cycle property, and acknowledged its roots in Axiomatic Design. In the following year, the book *Axiomatic Design in Large Systems* (2016) used the term ***Hetero-functional network*** for the first time. The first chapter explicitly links the Axiomatic Design literature to hetero-functional graph theory and engineering systems. Since that time, applications of hetero-functional graph theory have continued to expand. Prof. Khayal's research has applied hetero-functional graph theory to personalized healthcare delivery systems and W.C.H. Schoonenberg has sought to integrate these efforts into interdependent smart city infrastructures.

The Goal of This Book

Consequently, the goal of this book is to present, in one volume, a consistent hetero-functional graph theoretic treatment of interdependent smart city infrastructures as an ***overarching*** application domain of engineering systems. Naturally, in doing so, the work seeks to reconcile over a decade of research, including the many enhancements that came from tackling new and exciting application domains. Over the course of the text, we have made every effort to provide historical footnotes of how the theory has developed over that time. We have many hopes for the broad appeal of this work. To the systems engineering community, we hope that hetero-functional graph theory will be accepted as a quantification of many of the structural concepts found in model-based systems engineering languages like SysML. To the network science community, we hope to present a new view as how to construct graphs with fundamentally different meaning and insight. Finally, it is our hope that hetero-functional graph theory serves to overcome many of the theoretical and modeling limitations that have hindered our ability to systematically understand the structure and function of smart cities.

What Is in This Book?

This book is organized into seven chapters:

- Chapter 1 introduces the work in terms of the practical need to address smart cities as a pressing grand challenge. It also identifies the original contributions of the book and outlines how its argument evolves in the following chapters.

- Chapter 2 then turns to present the theoretical need for hetero-functional graph theory. An extensive discussion of multi-layer networks is provided. Its limitations are identified by means of a simple example of a hypothetical four-layer network.
- Chapter 3 orients the reader with hetero-functional graph theory preliminaries. The fundamental ontological concepts of soundness, completeness, lucidity, and laconicity are presented as means by which to formally assess the fidelity of a model. Multi-layer networks are found to lack completeness and lucidity. The remainder of the chapter relates the systems engineering foundations for hetero-functional graph theory. In particular, it focuses on the concept of system architecture at the instantiated, reference, and meta levels of abstraction.
- Chapter 4 relates hetero-functional graph theory rigorously as an intellectual fusion of model-based systems engineering and graph theory in terms of its seven constituent mathematical models. Simple examples are provided for each of these so as to demonstrate the conceptual links with SysML as a model-based systems engineering language. Formal definitions of all concepts are also provided so as to facilitate an explicit discussion of the underlying ontological structure.
- Chapter 5 then applies hetero-functional graph theory to an interdependent smart city infrastructure test case called "*Trimetrica*." One feature of "*Trimetrica*" is its significant heterogeneity of function. The chapter demonstrates the construction of a single system adjacency matrix for such a heterogeneous system. It subsequently discusses how this demonstration overcomes many limitations found in the multi-layer network literature.
- Chapter 6 serves to point the reader to further applications of hetero-functional graph theory. In particular, it summarizes its contributions to (1) mass-customized production systems, (2) transportation systems, (3) electric power systems, (4) electrified transportation systems, (5) microgrid-enabled production systems, and (6) personalized healthcare delivery systems. Along the way, the chapter highlights how hetero-functional graph theory can be used to create dynamical system simulation models and study life-cycle properties.
- Chapter 7 brings the book to a conclusion. It discusses some fertile areas for future research including the quantitative understanding of life-cycle properties, the treatment of cyber-physical systems, and the application of the network science literature on hetero-functional graphs.

In all, these seven chapters provide the reader with a rigorous introduction to hetero-functional graph theory so as to begin making independent contributions to the literature.

Hanover, NH, USA — Wester C. H. Schoonenberg
Lebanon, NH, USA — Inas S. Khayal
Hanover, NH, USA — Amro M. Farid
June 2018

Contents

List of Figures

List of Tables

Nomenclature

Ontological Symbols

$\mathcal{A}$	Abstraction, equals the mental conceptualization of the modeller.	23
$\mathcal{C}$	Domain Conceptualization, equals the understanding of reality.	23
$\mathcal{D}$	Real Domain, equals reality.	23
$\mathcal{L}$	Language, equals the description of reality.	23
$\mathcal{M}$	Model, equals the description of the abstraction.	23

Sets

$\overline{R}$	Set of Aggregated Resources	49
$\overline{R}_P$	Set of Aggregated Physical Resources	63
B	Set of Independent Buffers	40
B_S	Set of Buffers	43
H	Set of Transportation Resources	40
L	Set of Services	70
M	Set of Transformation Resources	40
M_{l_i}	Set of arcs: (service states to service activities) and (service activities to service states)	71
P	Set of System Processes	43
P_η	Set of Transportation Processes	43
P_μ	Set of Transformation Processes	43
P_Q	Set of Decision Algorithms	61
$P_{\bar{\eta}}$	Set of Refined Transportation Processes	44
P_γ	Set of Holding Processes	44
Q	Set of Cyber-Resources	60
Q_D	Set of Dependent Cyber-Resources	61
Q_I	Set of Independent Cyber-Resources	61
R	Set of System Resources	40
R_P	Set of Physical Resources	60
S_{l_i}	Set of Places describing the service states for product l_i	71
W_{l_i}	Set of Weights on the arcs describing the probabilities	71
$\mathcal{E}$	Set of System Actions	48

$\mathcal{E}_S$	Set of Structural Degrees of Freedom	52
$\mathcal{E}_{l_i}$	Set of Service Activities in Service l_i	70
$\mathcal{E}_{LH}$	Set of Service Transportation Degrees of Freedom	79
$\mathcal{E}_{LM}$	Set of Service Transformation Degrees of Freedom	79
$\mathcal{E}_{LS}$	Set of Service Degrees of Freedom	79
$\mathcal{Z}$	Set of System Activity Strings	56

Set Elements

b	Independent buffer in set of independent buffers B	40
b_{sy_1}	Origin buffer y_1 in set of buffers B_S	44
b_{sy_2}	Destination buffer y_2 in set of buffers B_S	44
e_{wv}	System action w, v in set of system actions $\mathcal{E}$	48
e_{xl_i}	Service activity x in Service l_i	70
h	Transporter in set of transporters H	40
l_i	Service i in the set of services L	70
m	Machine in set of machines M	40
p_Q	Decision algorithm in the set of decision algorithms P_Q	61
$p_{\bar{\eta}\varphi}$	Refined transportation process φ in set of refined transportation processes $P_{\bar{\eta}}$	45
$p_{\eta u}$	Transportation process u in set of transportation processes P_η	44
$p_{\gamma g}$	Holding process g in set of holding processes P_γ	44
$p_{\mu j}$	Transformation process j in set of transformation processes P_μ	43
q	Cyber-resource in the set of cyber-resources Q	60
r	Resource in set of system resources R	40
z_{χ_1,χ_2}	String of two sequential activities χ_1 and χ_2 in set of strings $\mathcal{Z}$	56

Indices

χ	Index in $[1\ldots\sigma(R)\sigma(P)]$	56
ψ	Index of the elements in the set of structural degrees of freedom $\mathcal{E}_S$	58
φ	Index of refined transportation process $p_{\bar{\eta}\varphi}$ in $P_{\bar{\eta}}$	45
g	Index of holding process $p_{\gamma g}$ in set of holding processes P_γ	44
i	Index of Service l_i in the set of services L	70
j	Index of transformation process $p_{\mu j}$ in set of transformation processes P_μ	43
k	Index of time	71
u	Index of transportation process $p_{\eta u}$ in set of transportation processes P_η	44
v	Index of physical resource p_w in set of physical resources R	48
w	Index of system process p_w in set of system processes P	48
x	Index of service activity e_{xl_i} in the set of service activities $\mathcal{E}_{l_i}$	70
y_1	Index of origin buffer b_{sy_1} in set of buffers B_S	44
y_2	Index of destination buffer b_{sy_2} in set of buffers B_S	44

Mathematical Symbols

$\mathbb{1}^n$	Ones-vector of length n	49
$\mathbb{A}$	System Adjacency Matrix	82

$\mathbb{A}_L$	System Service Adjacency Matrix	82
$\mathbb{A}_{\rho C}$	System Controller Agency Matrix	82
$\mathbb{A}_{\rho L}$	Service System Feasibility Matrix	82
$\mathbb{A}_{C\rho}$	Controller System Agency Matrix	82
$\mathbb{A}_{L\rho}$	System Service Feasibility Matrix	82
$\mathbb{P}_S$	A (non-unique) projection matrix for the vectorized knowledge base	58
$\overline{A}_Q$	Independent Controller Agency Matrix that shows jurisdiction of Q_I over R_P	63
$\Lambda_{\gamma i}$	Service Transportation Feasibility Matrix for product l_i	76
$\Lambda_{\mu i}$	Service Transformation Feasibility Matrix for product l_i	76
Λ_{Hi}	Transportation service selector matrix	79
Λ_{HL}	Transportation service line selector matrix	79
Λ_{Hxi}	Transportation service activity selector matrix	79
Λ_i	Service Feasibility Matrix for product l_i	77
Λ_{Mi}	Transformation service selector matrix	79
Λ_{ML}	Transformation service line selector matrix	79
Λ_{Mxi}	Transformation service activity selector matrix	79
Λ_{SHi}	System Transportation Service selector matrix	79
Λ_{SHxi}	System Transportation service activity selector matrix	79
Λ_{Si}	System Transformation Service selector matrix	79
Λ_{SL}	System Service Line selector matrix	79
Λ_{SMxi}	System Transformation service activity selector matrix	79
Λ_{Sxi}	System Transformation and Transportation service activity selector matrix	79
Φ_T	State Transition Function for a timed Petri net	71
Ξ	Resource Aggregation Matrix	49
$\widetilde{A}_\rho$	Hetero-functional Adjacency Matrix after elimination of row and column sparsity	58
A_C	Controller Adjacency Matrix	67
A_Q	Controller Agency Matrix	63
A_S	System Concept	52
A_ρ	Hetero-functional Adjacency Matrix	55
DOF_ρ	Sequence-Dependent Degrees of Freedom	58
DOF_H	Transportation Degrees of Freedom	52
DOF_M	Transformation Degrees of Freedom	52
DOF_S	Structural Degrees of Freedom	52
$DOF_{HH\rho}$	Measure of Type IV Sequence-Dependent Production Degrees of Freedom	56
$DOF_{HM\rho}$	Measure of Type III Sequence-Dependent Production Degrees of Freedom	56
DOF_{LH}	The number of transportation capabilities utilized by all services	80
DOF_{LM}	The number of transformation capabilities utilized by all services	80
DOF_{LS}	The number of capabilities utilized by all services	80
$DOF_{MH\rho}$	Measure of Type II Sequence-Dependent Production Degrees of Freedom	56

$DOF_{MM\rho}$	Measure of Type I Sequence-Dependent Production Degrees of Freedom	56
I^n	Identity Matrix of size $n \times n$	63
J_γ	Holding Knowledge Base	49
J_H	Transportation Knowledge Base	49
J_M	Transformation Knowledge Base	49
J_S	System Knowledge Base	48
$J_{\bar{H}}$	Refined Transportation Knowledge Base	49
J_ρ	System Sequence Knowledge Base	55
$J_{HH\rho}$	Type IV Sequence-Dependent Knowledge Base	56
$J_{HM\rho}$	Type III Sequence-Dependent Knowledge Base	56
$J_{MH\rho}$	Type II Sequence-Dependent Knowledge Base	56
$J_{MM\rho}$	Type I Sequence-Dependent Knowledge Base	56
K_ρ	System Sequence Constraints Matrix	55
K_M	Transformation Constraints Matrix	51
K_S	System Constraints Matrix	51
$K_{\bar{H}}$	Refined Transportation Constraints Matrix	51
$K_{HH\rho}$	Type IV Sequence-Dependent Constraints Matrix	56
$K_{HM\rho}$	Type III Sequence-Dependent Constraints Matrix	56
$K_{MH\rho}$	Type II Sequence-Dependent Constraints Matrix	56
$K_{MM\rho}$	Type I Sequence-Dependent Constraints Matrix	56
N_{l_i}	Service Net for product l_i	71
$Q_{El_i}[k]$	Marking of Service Transitions for product l_i at time k	71
Q_{l_i}	Petri net marking representing the set of service states	71
$Q_{Sl_i}[k]$	Marking of Service States for product l_i at time k	71
$U^+_{l_i}[k]$	Binary Input Firing Vector for product l_i at time k	71
$U^-_{l_i}[k]$	Binary Output Firing Vector for product l_i at time k	71
z_{l_i}	Sequence of Service Activities $e_{xl_i} \forall x \in [1, \ldots, \sigma(\mathcal{E}_{l_i})]$	71
$\mathcal{F}_v$	Resource Flexibility	49
$\mathcal{R}_w$	Process Redundancy	48

Mathematical Operators

$()^V$	Shorthand for vectorization (i.e. vec())	56
$\circledast$	Matrix Aggregation Operator	49
$\langle A, B \rangle_F$	Frobenius Product of matrices A and B	52
$\odot$	Matrix boolean multiplication	48
$\otimes$	Kronecker Product	49
$\sigma()$	The size of the set ().	44
$\times$	Cartesian Product	45

Chapter 1
Introduction

Cities have always played a prominent role in the prosperity of civilization. Indeed, every great civilization we can think of is associated with the prominence of one or more thriving cities. And so understanding cities—their inhabitants, their institutions, their infrastructure—what they are and how they work independently and together—is of fundamental importance to our collective growth as a human civilization. But the twenty-first century city will emerge as a societal grand challenge [136]. Two megatrends force cities to re-architect their infrastructure, society, and governance [85]. First, the world's population is growing and increasingly moving to urban areas. Currently, 55% of people live in cities that cover 3% of the earth [137], and another 2.5 billion are expected to move to cities by 2050 [136]. Second, climate change will impact urban areas disproportionally, whether as a consequence of extreme weather events or sea-level rise [14, 51], as the majority of megacities is located in coastal areas [105]. As a consequence of the two megatrends, cities will become more crowded, stressing the existing infrastructure, while they are increasingly threatened by the impact of climate change. Smart cities offer a solution, as they promise to create a more resilient and sustainable city [102].

The topic of smart cities has evolved over the past four decades, starting as early as the late 1970s with the concept of the "wired city" [21–24, 30, 59, 90]. More recently, in the past decade, research interest has spiked and a number of prominent smart city reviews have been published [6, 26, 29, 35, 62, 63, 81, 103, 125, 135, 147]. Despite these extensive efforts, a literature consensus has yet to form around a single smart city definition. Instead, Albino et al. [5] list 23 definitions in their review. Nevertheless, the smart city literature can be understood to take three interrelated [7], but distinct research thrusts [34, 99, 100]:

1. **(Interdependent) Infrastructure Systems** [87, 104]: These include all 16 critical infrastructure systems defined within the Presidential Policy Directive on Critical Infrastructure Resilience and Security [134].

W. C. H. Schoonenberg et al., *A Hetero-functional Graph Theory for Modeling Interdependent Smart City Infrastructure*,
https://doi.org/10.1007/978-3-319-99301-0_1

2. **People and Society** [55, 82, 89, 140]: These include desirable societal characteristics, structures, and "utopian" visions.
3. **Governance and Institutions** [8, 9, 20, 64, 95, 126, 139]: These include discussions on the enhancement of policy and regulations in the context of top-down vs. bottom-up approaches to empowerment and participation.

This book focuses on the first of these while explicitly recognizing the importance of the other two.

The emergence of smart cities is driven by the development of a "Future Internet" that consists of the Internet of Things, the Internet of Services, and the Internet of People [62] and acts as a platform to connect the smart city's devices and sensors. Although there are significant technical challenges associated with competing non-interoperable platforms [3, 31, 56, 70, 110, 113, 121, 130, 150], over time, these are likely to converge so as to have ubiquitous data throughout the city. While the explosion of data pertaining to smart city activities can bring innumerable benefits, there remain open and unresolved concerns around the value for privacy in society as much smart city data (e.g., healthcare information) is critically sensitive [40, 149]. The emergence of data from ubiquitous sensors will also give rise to big data analytics for data aggregation, processing, and interpretation [56, 146]. While these big data analytics efforts are well on their way [1, 76–78, 80, 114, 127], some have recognized the need for theory development to guide these analytical processes [18–20, 143].

The strong drive towards big data analytics lends itself to immediate case study conclusions about certain cities. Several case studies have been reported [25, 32, 83, 109, 131]. There is, however, a real doubt as to whether these conclusions are broadly applicable from city to city [16, 50, 142], either in the scope of their data or in the nature of their analytical methods. This book recognizes the need to develop a smart city theoretical literature to address the tremendous heterogeneity of cities so as to understand the applicability of case specific conclusions to other cities around the world [10–12].

Indeed, the continual progress in big data analytics forces the smart cities community to develop new theory that facilitates the planning and architecting of cities [18, 20]. For example, several efforts have been made to compare the "performance" of smart cities across the world [5, 53, 88, 91, 93]. Other works aim to define a theory for smart cities [58, 72, 84, 86, 92, 145]. These works remain qualitative in nature and have yet to develop sufficient breadth to accommodate the heterogeneity of smart cities. As smart cities become more crowded, systems theory [17] and design theory [129] recognize that a large number of functional requirements imposed on a confined space (the city) inherently couple the elements of the systems. Consequently, the infrastructures become interdependent [33, 38, 41, 61, 108, 115–117, 119].

The interdependence of smart city infrastructures raises timely and poignant questions about resilience and sustainability [87, 104]. No longer can cities simply focus on the cost-effective and reliable delivery of services. Such considerations must be completed by a concerted effort to develop life-cycle properties such as resilience and sustainability [39]. For example, there is a growing recognition that in order to achieve long-term sustainability through deep decarbonization, cities will have to invest heavily in electrified transportation [44, 45, 69, 138]. And yet, the experience of hurricane Sandy informs us that it is impossible to evacuate a city as large as New York City without the electrified subway and commuter rail systems. In other words, whereas electrified transportation can serve to bring about greater sustainability, it may indeed hinder a cities' resilient response to natural disasters (which may indeed be more frequent due to global climate change).

Thus far, the systems literature has developed many measures of resilience and sustainability [71, 101, 106, 107, 124, 132, 133, 148, 151, 152]. Many of these rely on a graph theoretic framework [4, 13, 15, 27, 36, 52, 54, 57, 60, 65, 66, 68, 96–98, 118, 120, 128, 144]. In such cases, the graph defined as a tuple of identical nodes and identical edges that connect them is well-suited to the study of a single homogeneous infrastructure [13, 28, 36, 49, 67]. In contrast, the assumption of a set of identical nodes and identical edges is ill-suited to an interdependent infrastructure system composed of two infrastructure networks of fundamentally different function. To overcome this limitation, the multi-layer network [37, 79] has developed. However, as Chapter 2 describes in detail, there remains significant modeling limitations to this body of work.

1.1 Book Contribution

This book advances a Hetero-functional Graph Theory (HFGT) to analyze the structure of interdependent smart city infrastructures. This work may be contrasted to the growing literature on multi-layer networks [37, 79], which is discussed extensively in Chapter 2. While significant theoretical advances in multi-layer networks have been made over the last decade, there is a recognition that multi-layer networks have modeling limitations [79] that prevent their real-world application to interdependent smart city infrastructures of arbitrary topology. In contrast, this book demonstrates that hetero-functional graph theory can be applied to arbitrary topologies of interdependent smart city infrastructures. The theory is extensible in the number of physical elements and functions within the city. It can also accommodate as many infrastructure systems (or layers) as required in the analysis. As discussed in the preface, this book does not constitute the first work on hetero-functional graph theory. The theory originates from the automated mass-customized production system literature where production systems were assumed to have arbitrary size and unlimited diversity of production capabilities for the manufacture of a nearly infinite number of product variants [94, 112]. For example, it is well-known that a 2002 Mercedes-Benz E-class can be produced in over 3.3×10^{24} different configurations [111].

The first conference paper, journal paper, and doctoral thesis on hetero-functional graph theory were published between 2006–2008 [42, 47, 48]. Since then, the theory has matured in the context of production systems, and has also been reapplied to several new application domains, including transportation systems [141], electric power systems [43], water distribution systems, and healthcare delivery systems [73–75]. Furthermore, the theory has been applied to application domains that consist of two interdependent infrastructure systems including Electrified Transportation systems [45, 46, 138], Microgrid-Enabled Production Systems [122, 123], and Energy-Water Nexus [2]. This book generalizes these works and shows how the theory may be applied to a system with an arbitrary number of interdependent infrastructures.

This book also contributes to the hetero-functional graph theory literature. It integrates, for the first time, all six matrices of hetero-functional graph theory in a single system adjacency matrix. This book also explicitly discusses how to characterize an interdependent infrastructure system composed of an *arbitrary number* of discipline-specific infrastructure. Consequently, it is the first work to demonstrate hetero-functional graph theory on a system with three interdependent infrastructures: a water distribution system, an electric power system, and a transportation system. The work shows that this seemingly incremental extension is of permutation complexity. Whereas two infrastructure systems can interact in $P(2,2) = 2!/0! = 2$ ways, three infrastructure systems can interact in $P(3,2) = 3!/1! = 6$ ways. The ability of hetero-functional graph theory to handle this permutation complexity speaks to its broad versatility.

This book also grounds hetero-functional graph theory to several existing streams in the systems literature. As expected, hetero-functional graph theory is presented in a language germane to the graph theory and the network science literature. The book also demonstrates that hetero-functional graph theory can be viewed as a quantification of many of the structural concepts found in model-based system engineering and its associated languages (e.g., SysML). Consequently, the book also draws on ontological science to discuss the modeling fidelity of hetero-functional graph theory relative to the existing multi-layer network literature. In other words, in developing a theory like hetero-functional graph theory, it is important to assess not just whether or not it possesses enough formal constructs to address the conceptualization of the system but also whether these formal concepts map one-to-one to the conceptualization. To our knowledge, this type of analysis is unprecedented in the network science literature.

The book makes every effort to be accessible to a broad audience of infrastructure system practitioners and researchers (e.g., electric power system planners, transportation engineers, and hydrologists, etc.). Consequently, the book has extensively visualized the graph theoretic concepts for greater intuition and clarity. Nevertheless, the book does require a common methodological base of its readers. Here, the book directs itself to the Model-Based Systems Engineering (MBSE) community and the Network Science Community (NSC). The traditional concepts from these two communities are assumed as an entry point to hetero-functional graph theory as a whole.

1.2 Book Outline

The book is organized as follows:

- Chapter 2 discusses the need for hetero-functional graph theory by demonstration of the limitations of existing multi-layer network theory.
- Chapter 3 introduces ontological science, model-based system engineering and graph theory as a background for hetero-functional graph theory.
- Chapter 4 then provides *Hetero-functional Graph Theory* as a theory that adheres to the ontological properties, and provides a discussion of its modeling elements.
- Chapter 5 introduces the "Trimetrica" test case as an interdependent smart city infrastructure system and then uses hetero-functional graph theory to describe the test case as a single mathematical model.
- Chapter 6 describes further applications of hetero-functional graph theory.
- Finally, Chapter 7 concludes the book and describes several promising directions for future work.

References

1. Abdel-Basset, M., & Mohamed, M. (2018). The role of single valued neutrosophic sets and rough sets in smart city: Imperfect and incomplete information systems. *Measurement, 124*, 47–55.
2. Abdulla, H., & Farid, A. M. (2015). Extending the energy-water nexus reference architecture to the sustainable development of agriculture, industry & commerce. In *First IEEE International Smart Cities Conference* (pp. 1–7). Guadalajara, 2015. http://dx.doi.org/10.1109/ISC2.2015.7366166
3. Abu-Matar, M., & Davies, J. (2018). Data driven reference architecture for smart city ecosystems. arXiv preprint arXiv:1805.01120.
4. Albert, R., Jeong, H., & Barabási, A.-L. (2000). Error and attack tolerance of complex networks. *Nature, 406*(6794), 378–382.
5. Albino, V., Berardi, U., & Dangelico, R. M. (2015). Smart cities: Definitions, dimensions, performance, and initiatives. *Journal of Urban Technology, 22*(1), 3–21.
6. Allwinkle, S., & Cruickshank, P. (2011). Creating smart-er cities: An overview. *Journal of Urban Technology, 18*(2), 1–16.
7. Andrisano, O., Bartolini, I., Bellavista, P., Boeri, A., Bononi, L., Borghetti, A., et al. (2018). The need of multidisciplinary approaches and engineering tools for the development and implementation of the smart city paradigm. *Proceedings of the IEEE, 106*(4), 738–760.
8. Angelidou, M. (2014). Smart city policies: A spatial approach. *Cities, 41*, S3–S11.
9. Angelidou, M. (2015). Smart cities: A conjuncture of four forces. *Cities, 47*, 95–106.
10. Angelidou, M. (2017). Shortcomings to smart city planning and development - exploring patterns and relationships. *Tema. Journal of Land Use, Mobility and Environment, 10*(1), 77–94.
11. Angelidou, M. (2017). The role of smart city characteristics in the plans of fifteen cities. *Journal of Urban Technology, 24*(4), 3–28.
12. Anthopoulos, L. (2017). Smart utopia vs smart reality: Learning by experience from 10 smart city cases. *Cities, 63*, 128–148.
13. Ash, J., & Newth, D. (2007). Optimizing complex networks for resilience against cascading failure. *Physica A: Statistical Mechanics and Its Applications, 380*, 673–683 http://www.sciencedirect.com/science/article/pii/S0378437107002543

14. Balk, D., Montgomery, M. R., McGranahan, G., Kim, D., Mara, V., Todd, M., et al. (2009). Mapping urban settlements and the risks of climate change in Africa, Asia and South America. In *Population dynamics and climate change* (Vol. 80). London: International Institute for Environment and Development.
15. Barker, K., Ramirez-Marquez, J. E., & Rocco, C. M. (2013). Resilience-based network component importance measures. *Reliability Engineering and System Safety, 117*, 89–97.
16. Bates, O., & Friday, A. (2017). Beyond data in the smart city: Repurposing existing campus IoT. *IEEE Pervasive Computing, 16*(2), 54–60.
17. Batty, M. (2008). The size, scale, and shape of cities. *Science, 319*(5864), 769–771.
18. Batty, M. (2012). Smart cities, big data. *Environment and planning B: Planning and Design, 39*, 191–193.
19. Batty, M. (2013). Big data, smart cities and city planning. *Dialogues in Human Geography, 3*(3), 274–279.
20. Batty, M., Axhausen, K. W., Giannotti, F., Pozdnoukhov, A., Bazzani, A., Wachowicz, M., et al. (2012). Smart cities of the future. *The European Physical Journal Special Topics, 214*(1), 481–518.
21. Batty, M., Couclelis, H., & Eichen, M. (1997). Urban systems as cellular automata.
22. Batty, M., & Longley, P. A. (1994). *Fractal cities: A geometry of form and function*. New York: Academic.
23. Batty, M., & Xie, Y. (1994). From cells to cities. *Environment and Planning B: Planning and Design, 21*(7), S31–S48.
24. Batty, M., Xie, Y., & Sun Z. (1999). Modeling urban dynamics through GIS-based cellular automata. *Computers, Environment and Urban Systems, 23*(3), 205–233.
25. Bhati, A., Hansen, M., & Chan, C. M. (2017). Energy conservation through smart homes in a smart city: A lesson for Singapore households. *Energy Policy, 104*, 230–239.
26. Bibri, S. E., & Krogstie, J. (2017) Smart sustainable cities of the future: An extensive interdisciplinary literature review. *Sustainable Cities and Society, 31*, 183–212.
27. Buldyrev, S. V., Parshani, R., Paul, G., Stanley, H. E., & Havlin, S. (2010). Catastrophic cascade of failures in interdependent networks. *Nature, 464*(7291), 1025–1028.
28. Callaway, D. S., Newman, M. E. J., Strogatz, S. H., & Watts, D. J. (2000). Network robustness and fragility: Percolation on random graphs. *Physical Review Letters, 85*, 5468–5471. http://link.aps.org/doi/10.1103/PhysRevLett.85.5468
29. Caragliu, A., Del Bo, C., & Nijkamp, P. (2011). Smart cities in Europe. *Journal of Urban Technology, 18*(2), 65–82.
30. Carter, N. (1995). MFP Australia: A vision of sustainable development for a post-industrial society. *Planning Practice & Research, 10*(1), 25–44.
31. Centenaro, M., Vangelista, L., Zanella, A., & Zorzi, M. (2016). Long-range communications in unlicensed bands: The rising stars in the IoT and smart city scenarios. *IEEE Wireless Communications, 23*(5), 60–67.
32. Chang, A. S., & Kalawsky, R. S. (2017). European transport sector intervention for smart city. In *2017 7th International Conference on Power Electronics Systems and Applications: Smart Mobility, Power Transfer & Security (PESA 2017)*. New York: IEEE. https://dspace.lboro.ac.uk/2134/27617
33. Cheng, J., Liu, Q., Hui, Q., & Choobineh, F. (2018). The joint optimization of critical interdependent infrastructure of an electricity-water-gas system. arXiv preprint arXiv:1803.02692.
34. Chourabi, H., Nam, T., Walker, S., Gil-Garcia, J. R., Mellouli, S., Nahon, K., et al. (2012). Understanding smart cities: An integrative framework. In: *2012 45th Hawaii International Conference on System Science (HICSS)* (pp. 2289–2297). New York: IEEE.
35. Cocchia, A. (2014). In C. R.-S. Renata Paola Dameri (Ed.), *Smart and digital city: A systematic literature review*. New York: Springer.
36. Colbourn, C. J. (1987). Network resilience. *SIAM. Journal on Algebraic and Discrete Methods, 8*(3), 404–409.
37. De Domenico, M., Solé-Ribalta, A., Cozzo, E., Kivelä, M., Moreno, Y., Porter, M. A., et al. (2013). Mathematical formulation of multilayer networks. *Physical Review X, 3*(4), 041022

38. De Porcellinis, S., Setola, R., Panzieri, S., & Ulivi, G. (2008). Simulation of heterogeneous and interdependent critical infrastructures. *International Journal of Critical Infrastructures, 4*(1–2), 110–128.
39. De Weck, O. L., Roos, D., & Magee, C. L. (2011). *Engineering systems: meeting human needs in a complex technological world*. Cambridge, MA: MIT Press. http://www.knovel.com/knovel2/Toc.jsp?BookID=4611http://mitpress-ebooks.mit.edu/product/engineering-systems
40. Eckhoff, D., & Wagner, I. (2017). Privacy in the smart city–applications, technologies, challenges and solutions. *IEEE Communications Surveys & Tutorials, 20*, 489–516.
41. Eusgeld, I., Nan, C., & Dietz, S. (2011). System-of-systems approach for interdependent critical infrastructures. *Reliability Engineering & System Safety, 96*(6), 679–686.
42. Farid, A. M. (2007). *Reconfigurability Measurement in Automated Manufacturing Systems*. Ph.D. Dissertation, University of Cambridge Engineering Department Institute for Manufacturing. http://engineering.dartmouth.edu/liines/resources/Theses/IEM-TP00.pdf
43. Farid, A. M. (2015). Multi-agent system design principles for resilient coordination and control of future power systems. *Intelligent Industrial Systems, 1*(3), 255–269. http://dx.doi.org/10.1007/s40903-015-0013-x
44. Farid, A. M. (2015). Symmetrica: Test case for transportation electrification research. *Infrastructure Complexity, 2*(9), 1–10. http://dx.doi.org/10.1186/s40551-015-0012-9
45. Farid, A. M. (2016). A hybrid dynamic system model for multi-modal transportation electrification. *IEEE Transactions on Control System Technology, 25*(99), 1–12. http://dx.doi.org/10.1109/TCST.2016.2579602
46. Farid, A. M. (2017). Electrified transportation system performance: Conventional vs. online electric vehicles. In N. P. Suh & D. H. Cho (Eds.), *The on-line electric vehicle: Wireless electric ground transportation systems* (Chap. 20, pp. 279–313). Berlin: Springer. http://engineering.dartmouth.edu/liines/resources/Books/TES-BC05.pdf
47. Farid, A. M., & McFarlane, D. C. (2006). A development of degrees of freedom for manufacturing systems. In *IMS'2006: 5th International Symposium on Intelligent Manufacturing Systems: Agents and Virtual Worlds*, Sakarya (pp. 1–6). http://engineering.dartmouth.edu/liines/resources/Conferences/IEM-C02.pdf
48. Farid, A. M., & McFarlane, D. C. (2007). A design structure matrix based method for reconfigurability measurement of distributed manufacturing systems. *International Journal of Intelligent Control and Systems Special Issue – Invited Paper, 12*(2), 118–129. http://engineering.dartmouth.edu/liines/resources/Journals/IEM-J02.pdf
49. Farley, T. R., & Colbourn, C. J. (2009). Multiterminal measures for network reliability and resilience. In *7th International Workshop on Design of Reliable Communication Networks* (pp. 107–114).
50. Fernandez-Anez, V., Fernández-Güell, J. M., & Giffinger, R. (2017). Smart city implementation and discourses: An integrated conceptual model. the case of Vienna. *Cities, 78*, 4–16.
51. Field, C. B., Barros, V. R., Mach, K., & Mastrandrea, M. (2014). *Climate change 2014: Impacts, adaptation, and vulnerability* (Vol. 1). Cambridge and New York: Cambridge University Press.
52. Francis, R., & Bekera, B. (2014). A metric and frameworks for resilience analysis of engineered and infrastructure systems. *Reliability Engineering and System Safety, 121*, 90–103.
53. Girardi, P., & Temporelli, A. (2017). Smartainability: A methodology for assessing the sustainability of the smart city. *Energy Procedia, 111*, 810–816.
54. Gómez-Gardeñes, J., Reinares, I., Arenas, A., & Floría, L. M. (2012). Evolution of cooperation in multiplex networks. *Scientific Reports 2*, Article No 620.
55. Grossi, G., & Pianezzi, D. (2017). Smart cities: Utopia or neoliberal ideology? *Cities, 69*, 79–85.
56. Hancke, G. P., de Carvalho e Silva, B., & Hancke Jr., G. P. (2012). The role of advanced sensing in smart cities. *Sensors, 13*(1), 393–425.
57. Harary, F., & Hayes, J. P. (1993). Edge fault tolerance in graphs. *Networks, 23*(2), 135–142.

58. Harrison, C., & Donnelly, I. A. (2011). A theory of smart cities. In *Proceedings of the 55th Annual Meeting of the ISSS-2011, Hull* (Vol. 55, no. 1).
59. Heng, T. M., & Low, L. (1993). The intelligent city: Singapore achieving the next lap: Practitoners forum. *Technology Analysis & Strategic Management, 5*(2), 187–202.
60. Henry, D., & Emmanuel Ramirez-Marquez, J. (2012). Generic metrics and quantitative approaches for system resilience as a function of time. *Reliability Engineering and System Safety, 99*, 114–122.
61. Heracleous, C., Kolios, P., Panayiotou, C. G., Ellinas, G., & Polycarpou, M. M. (2017). Hybrid systems modeling for critical infrastructures interdependency analysis. *Reliability Engineering & System Safety,165*, 89–101.
62. Hernández-Muñoz, J. M., Vercher, J. B., Muñoz, L., Galache, J. A., Presser, M., Gómez, L. A. H., & Pettersson, J. (2011). Smart cities at the forefront of the future internet. In *The future internet assembly* (pp. 447–462). New York: Springer.
63. Hollands, R. G. (2008). Will the real smart city please stand up? *City, 12*(3), 303–320.
64. Hollands, R. G. (2015). Critical interventions into the corporate smart city. *Cambridge Journal of Regions, Economy and Society, 8*(1), 61–77.
65. Holme, P., Kim, B. J., Yoon, C. N., & Han, S. K. (2002). Attack vulnerability of complex networks. *Physical Review E, 65*(5), 1–14.
66. Hwang, F. K., Najjar, W., & Gaudiot, J. L. (1994). Comments on Network resilience: A measure of network fault tolerance [and reply]. *IEEE Transactions on Computers, 43*(12), 1451–1453. http://dx.doi.org/10.1109/12.338109
67. Hwang, F., Najjar, W., & Gaudiot, J. (1994). Comments on "network resilience: A measure of network fault tolerance" [with reply]". *IEEE Transactions on Computers, 43*(12), 1451–1453.
68. Ip, W. H., & Wang, D. (2011). Resilience and friability of transportation networks: Evaluation, analysis and optimization. *IEEE Systems Journal, 5*(2), 189–198.
69. Janić, M. (2014). *Advanced transport systems*. New York: Springer.
70. Jin, J., Gubbi, J., Marusic, S., & Palaniswami, M. (2014). An information framework for creating a smart city through internet of things. *IEEE Internet of Things Journal, 1*(2), 112–121.
71. Johansen, C., & Tien, I. (2018). Probabilistic multi-scale modeling of interdependencies between critical infrastructure systems for resilience. *Sustainable and Resilient Infrastructure, 3*(1), 1–15.
72. Joshi, S., Saxena, S., Godbole, T., & Shreya. (2016). Developing smart cities: An integrated framework. *Procedia Computer Science, 93*, 902–909.
73. Khayal, I. S., & Farid, A. M. (2015). Axiomatic design based volatility assessment of the Abu Dhabi healthcare labor market. *Journal of Enterprise Transformation, 5*(3), 162–191. http://dx.doi.org/10.1080/19488289.2015.1056449
74. Khayal, I. S., & Farid, A. M. (2016). The need for systems tools in the practice of clinical medicine. *Systems Engineering, 20*(1), 3–20. http://dx.doi.org/10.1002/sys.21374
75. Khayal, I. S., & Farid, A. M. (2017). Architecting a system model for personalized healthcare delivery and managed individual health outcomes. *Complexity, 1*(1), 1–25. https://doi.org/10.1155/2018/8457231
76. Kitchin, R. (2014). The real-time city? big data and smart urbanism. *GeoJournal, 79*(1), 1–14.
77. Kitchin, R. (2018). The realtimeness of smart cities. *TECNOSCIENZA: Italian Journal of Science & Technology Studies, 8*(2), 19–42.
78. Kitchin, R., & Dodge, M. (2017). The (in) security of smart cities: Vulnerabilities, risks, mitigation, and prevention. *Journal of Urban Technology*, 1–19 https://doi.org/10.1080/10630732.2017.1408002
79. Kivelä, M., Arenas, A., Barthelemy, M., Gleeson, J. P., Moreno, Y., & Porter, M. A. (2014). Multilayer networks. *Journal of Complex Networks, 2*(3), 203–271.
80. Kotevska, O., Kusne, A. G., Samarov, D. V., Lbath, A., & Battou A. (2017). Dynamic network model for smart city data-loss resilience case study: City-to-city network for crime analytics. *IEEE Access, 5*, 20524–20535.

81. Kummitha, R. K. R., & Crutzen, N. (2017). How do we understand smart cities? an evolutionary perspective. *Cities, 67*, 43–52.
82. Lacinák, M., & Ristvej, J. (2017). Smart city, safety and security. *Procedia Engineering, 192*, 522–527.
83. Lee, J. H., Hancock, M. G., & Hu, M.-C. (2014). Towards an effective framework for building smart cities: Lessons from Seoul and San Francisco. *Technological Forecasting and Social Change, 89*, 80–99.
84. Lee, J., & Lee, H. (2014). Developing and validating a citizen-centric typology for smart city services. *Government Information Quarterly, 31*, S93–S105.
85. Leichenko, R. (2011). Climate change and urban resilience. *Current opinion in environmental sustainability, 3*(3), 164–168.
86. Letaifa, S. B. (2015). How to strategize smart cities: Revealing the smart model. *Journal of Business Research, 68*(7), 1414–1419.
87. Linkov, I., & Palma-Oliveira, J. M. (2017). *Resilience and risk: Methods and application in environment, cyber and social domains*. New York: Springer.
88. Lombardi, P., Giordano, S., Farouh, H., & Yousef, W. (2012). Modelling the smart city performance. *Innovation: The European Journal of Social Science Research, 25*(2), 137–149.
89. March, H. (2016). The smart city and other ICT-led techno-imaginaries: Any room for dialogue with degrowth? *Journal of Cleaner Production, 197*, 1694–1703.
90. Martin, J. (1978). *The wired society*. Prentice-Hall: Upper Saddle River.
91. Martin, C. J., Evans, J., & Karvonen, A. (2018). Smart and sustainable? five tensions in the visions and practices of the smart-sustainable city in Europe and North America. *Technological Forecasting and Social Change, 133*, 269–278.
92. Mattoni, B., Gugliermetti, F., & Bisegna, F. (2015). A multilevel method to assess and design the renovation and integration of smart cities. *Sustainable Cities and Society, 15*, 105–119.
93. Maye, D. (2018). Smart food city': Conceptual relations between smart city planning, urban food systems and innovation theory. *City, Culture and Society*. https://doi.org/10.1016/j.ccs.2017.12.001
94. Mehrabi, M. G., Ulsoy, A. G., Koren, Y., & Heytler, P. (2002). Trends and perspectives in flexible and reconfigurable manufacturing systems. *Journal of Intelligent Manufacturing, 13*(2), 135–146.
95. Meijer, A., & Bolívar, M. P. R. (2016). Governing the smart city: a review of the literature on smart urban governance. *International Review of Administrative Sciences, 82*(2), 392–408.
96. Milanovic, J. V., & Zhu, W. (2017). Modelling of interconnected critical infrastructure systems using complex network theory. In *IEEE Transactions on Smart Grid*.
97. Morris, R. G., & Barthelemy, M. (2013). Interdependent networks: The fragility of control. *Scientific Reports 3* (2013).
98. Najjar, W., & Gaudiot, J.-L. (1990). Network resilience: A measure of network fault tolerance. *IEEE Transactions on Computers, 39*(2), 174–181.
99. Nam, T., & Pardo, T. A. (2011). Smart city as urban innovation: Focusing on management, policy, and context. In *Proceedings of the 5th International Conference on Theory and Practice of Electronic Governance* (pp. 185–194). New York: ACM.
100. Nam, T., & Pardo, T. A. (2011). Conceptualizing smart city with dimensions of technology, people, and institutions, In *Proceedings of the 12th Annual International Digital Government Research Conference: Digital Government Innovation in Challenging Times*. (ACM, New York, 2011), pp. 282–291
101. Nan, C., & Sansavini, G. (2017). A quantitative method for assessing resilience of interdependent infrastructures. *Reliability Engineering & System Safety, 157*, 35–53.
102. Naphade, M., Banavar, G., Harrison, C., Paraszczak, J., & Morris, R. (2011). Smarter cities and their innovation challenges. *Computer, 44*(6), 32–39.
103. Neirotti, P., De Marco A., Cagliano, A. C., Mangano, G., & Scorrano, F. (2014). Current trends in smart city initiatives: Some stylised facts. *Cities, 38*, 25–36.
104. Nelson, P. P., & Sterling, R. L. (2012). Sustainability and resilience of underground urban infrastructure: New approaches to metrics and formalism. In *GeoCongress 2012* (pp. 3199–3208), March 2012.

105. Neumann, B., Vafeidis, A. T., Zimmermann, J., & Nicholls, R. J. (2015). Future coastal population growth and exposure to sea-level rise and coastal flooding-a global assessment. *PloS one, 10*(3), e0118571.
106. Ouyang, M. (2017). A mathematical framework to optimize resilience of interdependent critical infrastructure systems under spatially localized attacks. *European Journal of Operational Research, 262*(3), 1072–1084.
107. Ouyang, M., & Wang, Z. (2015). Resilience assessment of interdependent infrastructure systems: With a focus on joint restoration modeling and analysis. *Reliability Engineering & System Safety, 141*, 74–82.
108. Pagani, G. A., & Aiello, M. (2013). The power grid as a complex network: A survey. *Physica A: Statistical Mechanics and its Applications, 392*(11), 2688–2700.
109. Paroutis, S., Bennett, M., & Heracleous, L. (2014). A strategic view on smart city technology: The case of IBM smarter cities during a recession. *Technological Forecasting and Social Change, 89*, 262–272.
110. Petrolo, R., Loscri, V., & Mitton, N. (2017). Towards a smart city based on cloud of things, a survey on the smart city vision and paradigms. *Transactions on Emerging Telecommunications Technologies, 28*(1). https://doi.org/10.1002/ett.2931
111. Pil, F. K., & Holweg, M. (2004). Linking product variety to order-fulfillment strategies. *Interfaces, 34*(5), 394–403.
112. Pine, J. B. (1993). *Mass customization: The new frontier in business competition*. Cambridge, MA: Harvard Business School Press.
113. Piro, G., Cianci, I., Grieco, L. A., Boggia, G., & Camarda, P. (2014). Information centric services in smart cities. *Journal of Systems and Software, 88*, 169–188.
114. Rathore, M. M., Ahmad, A., Paul, A., & Rho, S. (2016). Urban planning and building smart cities based on the internet of things using big data analytics. *Computer Networks, 101*, 63–80.
115. Rinaldi, S. M. (2004). Modeling and simulating critical infrastructures and their interdependencies. In *Proceedings of the 37th Annual Hawaii International Conference on System Sciences, 2004* (8 pp.) Jan 2004.
116. Rinaldi, S. M., Peerenboom, J. P., & Kelly, T. K. (2001). Identifying, understanding, and analyzing critical infrastructure interdependencies. *IEEE Control Systems, 21*(6), 11–25,
117. Rosato, V., Issacharoff, L., Tiriticco, F., Meloni, S., Porcellinis, S., & Setola, R. (2008). Modelling interdependent infrastructures using interacting dynamical models. *International Journal of Critical Infrastructures, 4*(1–2), 63–79. http://www.inderscienceonline.com/doi/abs/10.1504/IJCIS.2008.016092
118. Rosenkrantz, D. J., Goel, S., Ravi, S. S., & Gangolly, J. (2009). Resilience metrics for service-oriented networks: A service allocation approach. *IEEE Transactions on Services Computing, 2*(3), 183–196.
119. Saidi, S., Kattan, L., Jayasinghe, P., Hettiaratchi, P., & Taron, J. (2017). Integrated infrastructure systems—a review. *Sustainable Cities and Society, 36*, 1–11.
120. Salles, R. M., & Marino, D. A. (2011). Strategies and metric for resilience in computer networks. *The Computer Journal, 55*(6), 728–739.
121. Sanchez, L., Muñoz, L., Galache, J. A., Sotres, P., Santana, J. R., Gutierrez, V., et al. (2014). Smartsantander: IoT experimentation over a smart city testbed. *Computer Networks, 61*, 217–238.
122. Schoonenberg, W. C., & Farid, A. M. (2015). A dynamic production model for industrial systems energy management. In *2015 IEEE International Conference on Systems Man and Cybernetics*, Hong Kong, 2015 (pp. 1–7). http://dx.doi.org/10.1109/SMC.2015.14
123. Schoonenberg, W. C., & Farid, A. M. (2017). A dynamic model for the energy management of microgrid-enabled production systems. *Journal of Cleaner Production, 1*(1), 1–10.
124. Sen, A., Mazumder, A., Banerjee, J., Das, A., & Compton, R. (2014). Identification of K most vulnerable nodes in multi-layered network using a new model of interdependency. In *2014 IEEE Conference on Computer Communications Workshops (INFOCOM WKSHPS)* pp. (pp. 831–836). New York: IEEE.

125. Shapiro, J. M. (2006). Smart cities: Quality of life, productivity, and the growth effects of human capital. *The Review of Economics and Statistics, 88*(2), 324–335.
126. Shelton, T., Zook, M., & Wiig, A. (2015). The 'actually existing smart city'. *Cambridge Journal of Regions, Economy and Society, 8*(1), 13–25.
127. Sta, H. B. (2017). Quality and the efficiency of data in "smart-cities". *Future Generation Computer Systems, 74*, 409–416.
128. Sudakov, B., & Vu, V. H. (2008). Local resilience of graphs. *Random structures and algorithms* (pp. 409–433).
129. Suh, N. P. (2001). *Axiomatic design: Advances and applications*. Oxford: Oxford University Press.
130. Talari, S., Shafie-khah, M., Siano, P., Loia, V., Tommasetti, A., & Catalão, J. P. (2017). A review of smart cities based on the internet of things concept. *Energies, 10*(4), 421.
131. Tanda, A., De Marco, A., & Rosso, M. (2017). Evaluating the impact of smart city initiatives. In *Proceedings of the 6th International Conference on Smart Cities and Green ICT Systems (SMARTGREENS 2017)* (pp. 281–286). Scitepress Ed.
132. Thacker, S., Hall, J. W., & Pant, R. (2017). Preserving key topological and structural features in the synthesis of multilevel electricity networks for modeling of resilience and risk. *Journal of Infrastructure Systems, 24*(1), 04017043.
133. Thacker, S., Pant, R., & Hall J. W. (2017). System-of-systems formulation and disruption analysis for multi-scale critical national infrastructures. *Reliability Engineering & System Safety, 167*, 30–41.
134. The White House Office of the Press Secretary. (2013). *Presidential Policy Directive: Critical Infrastructure Security and Resilience (PPD-21)*. The White House, Washington, DC, Technical Report (2013). http://www.whitehouse.gov/the-press-office/2013/02/12/presidential-policy-directive-critical-infrastructure-security-and-resil
135. Trindade, E. P., Hinnig, M. P. F., da Costa, E. M., Marques, J. S., Bastos, R. C., & Yigitcanlar, T. (2017). Sustainable development of smart cities: a systematic review of the literature. *Journal of Open Innovation: Technology, Market, and Complexity, 3*(1), 11.
136. UN. (2015). UN sustainable development goals. 17 goals to transform our world, September 2015. https://www.un.org/sustainabledevelopment/sustainable-development-goals/
137. UN. (2018). 68% of the world population projected to live in urban areas by 2050, says UN, May 2018. https://www.un.org/sustainabledevelopment/blog/2018/05/68-world-population-projected-live-urban-areas-2050-says-un/
138. van der Wardt, T. J., & Farid, A. M. (2017). A hybrid dynamic system assessment methodology for multi-modal transportation-electrification. *Energies, 10*(5), p. 653. http://dx.doi.org/10.3390/en10050653
139. Vanolo, A. (2014). Smartmentality: The smart city as disciplinary strategy. *Urban Studies, 51*(5), 883–898.
140. Vanolo, A. (2016). Is there anybody out there? the place and role of citizens in tomorrow's smart cities. *Futures, 82*, 26–36.
141. Viswanath, A., Baca, E. E. S., & Farid, A. M. (2014). An axiomatic design approach to passenger itinerary enumeration in reconfigurable transportation systems. *IEEE Transactions on Intelligent Transportation Systems, 15*(3), 915–924. http://dx.doi.org/10.1109/TITS.2013.2293340
142. Walravens, N. (2015). Mobile city applications for brussels citizens: Smart city trends, challenges and a reality check. *Telematics and Informatics, 32*(2), 282–299.
143. West, G. (2013). Big data needs a big theory to go with it. https://www.scientificamerican.com/article/big-data-needs-big-theory/
144. Whitson, J. C., & Ramirez-Marquez, J. E. (2009). Resiliency as a component importance measure in network reliability. *Reliability Engineering & System Safety, 94*(10), 1685–1693. http://dx.doi.org/10.1016/j.ress.2009.05.001
145. Yamagata, Y., & Seya, H. (2013). Simulating a future smart city: An integrated land use-energy model. *Applied Energy, 112*, 1466–1474.

146. Yaqoob, I., Hashem, I. A. T., Gani, A., Mokhtar, S., Ahmed, E., Anuar, N. B., et al. (2016). Big data: From beginning to future. *International Journal of Information Management, 36*(6), 1231–1247
147. Yin, C., Xiong, Z., Chen, H., Wang, J., Cooper, D., & David, B. (2015). A literature survey on smart cities. *Science China Information Sciences, 58*(10), 1–18.
148. Zhang, C., Kong, J.-J., & Simonovic, S. P. (2018). Restoration resource allocation model for enhancing resilience of interdependent infrastructure systems. *Safety Science, 102*, 169–177.
149. Zhang, K., Ni, J., Yang, K., Liang, X., Ren, J., & Shen, X. S. (2017). Security and privacy in smart city applications: Challenges and solutions. *IEEE Communications Magazine, 55*(1), 122–129.
150. Zhuhadar, L., Thrasher, E., Marklin, S., & de Pablos, P. O. (2017). The next wave of innovation—review of smart cities intelligent operation systems. *Computers in Human Behavior, 66*, 273–281.
151. Zimmerman, R. (2004). Decision-making and the vulnerability of interdependent critical infrastructure. In *2004 IEEE International Conference on Systems, Man and Cybernetics* (Vol. 5, pp. 4059–4063). New York: IEEE.
152. Zuev, K., & Beer, M. (2017). Reliability of critical infrastructure networks: Challenges. arXiv preprint arXiv:1701.00594.

Chapter 2
The Need for Hetero-functional Graph Theory

Smart cities create the need for greater integration between multiple infrastructures. Here, it is important to note that the coupling between these multiple infrastructures may take on a wide variety of potential topologies. For example, a conventional road transportation system is coupled to the electric power system by lighting and sign posts [2]. The exchanged power flows are relatively small in comparison with the power consumed by either system independently. However, as electric vehicles, buses, and trains become a larger part of the transportation system, they introduce a much greater power coupling to the electric power system that may ultimately affect how both systems are designed and operated [40, 44, 50, 95]. Furthermore, charging stations and electrified rail introduce fundamentally different types of coupling. While the first couples a power system node to a transportation node, the latter couples it to transportation edge. Consequently, there is a need for a theoretic modeling foundation that recognizes the heterogeneity of multiple infrastructures coupled arbitrarily.

Because all infrastructures are spatially-distributed networks, their underlying models have often been rooted in graph theory [58, 69, 87]. This specifically includes conventional treatments of electric power systems [42], natural gas pipeline systems [26, 64], oil pipeline systems [26, 61], multi-modal transportation systems [24, 53, 67], as well as water distribution networks [37]. Naturally, as the study of interdependent critical infrastructures (ICIs) has advanced, many works have looked to graph theory and the network sciences [58, 69, 87] for this common modeling foundation.

Recently, the network science literature has advanced the concept of "multi-layer networks" where two or more network "layers" interact with each other to represent a system of ICIs [32, 46, 54]. Much like a conventional graph, a multi-layer network $G_M = \{V_M, E_M\}$ is formally defined as a tuple of nodes V_M and edges E_M. Such a multi-layer network is organized into an integer n number of layers $L_1 \dots L_n$. Here, a given layer $L_\alpha = \{V_\alpha, E_\alpha\}$ is understood as a graph where the nodes V_α and edges E_α have at least one semantic aspect, feature or operand in common (e.g., electricity,

W. C. H. Schoonenberg et al., *A Hetero-functional Graph Theory for Modeling Interdependent Smart City Infrastructure*,
https://doi.org/10.1007/978-3-319-99301-0_2

water, people, etc). Furthermore, the multi-layer network edges $E_M = E_A \cup E_C$ can be classified into intra-layer edges E_A and extra-layer edges E_C [54]. Despite these definitions, the multi-layer network community *"has produced an equally immense explosion of disparate terminology, and the lack of consensus (or even generally accepted) set of terminology and mathematical framework for studying is extremely problematic"* [54]. In a comprehensive review on the common ontological elements and modeling limitations of "multi-layer networks," Kivela et al. showed that *all* of the reviewed works have exhibited at least one of the following modeling constraints [54]:

1. Alignment of nodes between layers is required
2. Disjointment between layers is required
3. Equal number of nodes for all layers is required
4. Exclusively vertical coupling between all layers is required
5. Equal couplings between all layers are required
6. Node counterparts are coupled between all layers
7. Limited number of modeled layers
8. Limited number of aspects in a layer

The practical limitations of these modeling constraints are best explained graphically by counter-example on the small hypothetical four-layer network shown in Figure 2.1. This figure is adopted from Appendix A in which a more extensive description of the figure is added. The figure consists of three physical layers in green, yellow, and blue representing the transportation, electric power, and water distribution infrastructures, respectively. The fourth layer is the cyber layer (in grey) representing cyber-control infrastructure. Lastly, the network contains physical links between layers in dashed red and informatic links in dashed grey. In all, there are 16 nodes (or vertices), 9 intra-layer edges, and 15 cross-layer edges. Naturally, each node and edge coloring represents a different infrastructure system. Sometimes, nodes connected by cross-layer edges represent a single *resource* (or facility[1]) with presence in several infrastructures. For example, the group of nodes v_1, v_5, and v_8 can represent a house that combines a private parking lot in the transportation system (v_1), an electric power grid connection that supplies power to the house (v_5), and a water tap in the water distribution network (v_8). A resource r_i is, therefore, often defined by a graph $G_{Ri} = \{V_{Ri}, E_{Ri}\}$ composed of "counterpart" nodes V_{Ri} and coupling edges $E_{Ri} \subseteq E_C$. Arbitrary nodes v_i in layer L_α and v_j in layer L_β are said to be "counterparts" of each other if there exists no cross-layer edges between L_α and L_β involving v_i or v_j other than $e_{xk} = \{v_i, v_j\}$. That said, nodes do not need to be part of resources in order to be connected across layers. For example, v_{10} may represent a hot tub that is connected to the same electric load bus v_5, but draws its water from the water distribution system directly. Furthermore, Node v_3 may represent a charging station; as a resource that crosses the transportation and

[1]These resources or facilities are referred to as entities in Kivela et al. [54].

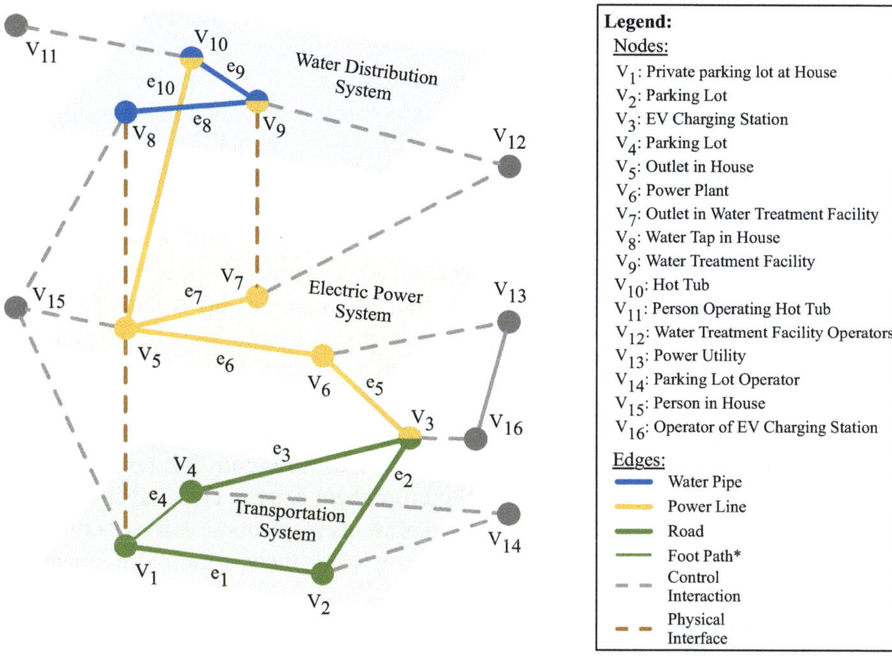

Fig. 2.1 A hypothetical four-layer network: it represents transportation, electric power, and water distribution infrastructure with a superimposed cyber-control layer. *: The foot path is part of the Transportation System, but differs in modality from the other edges in the system and is represented with a thinner edge

electricity infrastructures in a single node. Node v_{14} acts as a centralized controller agent for v_2 and v_4. It may represent a bus or taxi dispatching authority. Meanwhile, node v_{16} acts as another controller agent for node v_3. It may represent the charging station's cloud-based energy management software. Note that this topology may cause conflict between the transportation objective of node v_{14} and the energy management objective of node v_{16}. The remainder of this chapter uses Figure 2.1 to illustrate the eight modeling constraints found in multi-layer networks [54].

In *Constraint 1*, some "multi-layer networks" require all layers to have vertically aligned nodes [3–5, 7–17, 19, 20, 25, 27, 29, 29, 33, 34, 39, 43, 45, 47–49, 51, 52, 54, 55, 59, 62, 65, 66, 68, 68, 70, 71, 73, 74, 77–80, 85, 90, 91, 93]. In other words, for any node v_i in layer L_α, there must exist another *connected* node v_j in layer L_β. In Figure 2.1, however, v_4 is not connected to v_7. It, therefore, represents an example infrastructure with modeling requirements greater than those provided by multi-layer networks with vertically aligned nodes.

In *Constraint 2*, some "multi-layer networks" require disjoint layers [1, 6, 17, 18, 21, 35, 36, 47, 54, 55, 57, 60, 63, 72, 75, 76, 84, 88, 89, 96, 97]. In other words, any node v_i can only be part of a single layer L_α, furthermore, any two layers do not intersect. In Figure 2.1, however, the charging station, v_3, pertains to both the electrical layer and the transportation layer. Figure 2.1, therefore, represents a simple example infrastructure with modeling requirements greater than those provided by multi-layer networks with disjoint layers.

In *Constraint 3*, some "multi-layer networks" require the same number of nodes in each layer [4, 5, 7, 8, 10–16, 18–20, 27, 29, 33–35, 39, 41, 45, 47–49, 51, 52, 54, 56, 59, 60, 62, 63, 65, 66, 68, 70, 71, 73, 74, 77–80, 85, 91, 93]. In Figure 2.1, however, the water network has three nodes while the transportation network has four. Note that this constraint is similar to the first, except that it does not require for a node in a given layer to be connected to other layers. Figure 2.1, therefore, represents an example infrastructure with modeling requirements greater than those provided by multi-layer networks with an equal number of nodes in each layer.

In *Constraint 4*, some "multi-layer networks" require exclusively "vertical"[2] cross-layer couplings [4, 5, 7, 8, 10–16, 19–21, 27–30, 33, 34, 39, 45, 47–49, 51, 52, 54, 56, 59, 62, 65, 66, 68, 70, 71, 73, 74, 77–80, 85, 91–94]. In other words, all the interlayer edges $E_C = \cup_i E_{Ri}$ are coupling edges. In Figure 2.1, however, node v_{10} is not a counterpart of node v_5 and therefore there exists a cross-layer edge that is not a coupling edge. Again, Figure 2.1 represents an example infrastructure that violates a constraint imposed by multi-layer network theory; the requirement for "vertical" couplings.

In *Constraint 5*, some "multi-layer networks" require that all nodes in a given layer have identical couplings to nodes in another layer [4, 5, 8, 10–16, 19, 21, 27–30, 39, 45, 47–49, 51, 52, 54, 56, 59, 62, 65, 66, 70, 71, 73, 74, 77–80, 85, 91–94]. In Figure 2.1, however, node v_5 and v_8 have a cross-layer edge but v_4 and v_7 do not. Consequently, Figure 2.1 presents an example in which the requirement for identical cross-layer couplings is not met.

In *Constraint 6*, some "multi-layer networks" require that each node is connected to all of its counterparts in other layers [4, 5, 8, 10–16, 19, 21, 27, 28, 30, 39, 45, 47–49, 51, 52, 56, 59, 62, 65, 66, 70, 71, 73, 74, 79, 80, 85, 91–94]. In Figure 2.1, however, many nodes (e.g., v_2 and v_4) are not connected to other layers. Figure 2.1, therefore, represents an example infrastructure with modeling requirements greater than those provided by multi-layer networks with all nodes connected to their counterparts.

In *Constraint 7*, some "multi-layer networks" limit the number of layers to two [1, 6, 17, 18, 21–23, 28, 30, 31, 35, 36, 39, 41, 47, 55–57, 60, 62, 63, 65, 66, 72, 75, 76, 81–84, 86, 88, 89, 91, 92, 94, 96, 97]. In Figure 2.1, however, there are three physical networks and one cyber-network. Dual-layer analysis is likely insufficient in reflecting the number of layers, as is the case for this simple example. Figure 2.1, therefore, exceeds the maximum of two layers as imposed in Constraint 7 on any multi-layer network.

In *Constraint 8*, some "multi-layer networks" require that each layer have no more than one aspect [1, 4–8, 10–21, 27, 29, 33–36, 39, 45, 47–49, 51, 52, 55–

[2]Kivela et al. [54] refer to this constraint as "diagonal" couplings. This work adopts the term vertical to more closely reflect the depiction in Figure 2.1.

57, 59, 60, 62, 63, 65, 66, 68, 70–80, 84, 85, 88, 89, 91, 93, 96, 97]. However, many networks have multiple aspects. Transportation systems are often multi-modal including passenger vehicles, buses, trains, and pedestrians [38]. Figure 2.1 includes two modes of transportation, a foot path between nodes v_1 and v_4, and roads for all other connections. Therefore, the system represents an example where the modeling requirements exceed those provided by multi-layer networks with a single aspect.

These eight constraints show insufficiencies in the existing theory of multi-layer networks to address a simple example of multiple arbitrarily connected infrastructure systems. In other words, there exist ontological limitations to their associated models of the physical system. Consequently, a new theory needs to be introduced that does not have such limitations. Chapter 4 presents such a theory; herein called Hetero-functional Graph Theory (HFGT). It specifically addresses many of the limitations identified thus far in this chapter. In order to facilitate its explanations, the following Chapter 3 is dedicated to highlighting the fundamental concepts upon which HFGT is based.

References

1. Allard, A., Noël, P.-A., Dubé, L. J., & Pourbohloul, B. (2009). Heterogeneous bond percolation on multitype networks with an application to epidemic dynamics. *Physical Review E, 79*, 036113. http://link.aps.org/doi/10.1103/PhysRevE.79.036113
2. Atıcı, C., Ozcelebi, T., & Lukkien, J. J. (2011). Exploring user-centered intelligent road lighting design: A road map and future research directions. *IEEE Transactions on Consumer Electronics, 57*(2), 788–793.
3. Barigozzi, M., Fagiolo, G., & Garlaschelli, D. (2010). Multinetwork of international trade: A commodity-specific analysis. *Physical Review E, 81*, 046104. http://link.aps.org/doi/10.1103/PhysRevE.81.046104
4. Barigozzi, M., Fagiolo, G., & Mangioni, G. (2011). Identifying the community structure of the international-trade multi-network. *Physica A: Statistical Mechanics And Its Applications, 390*(11), 2051–2066.
5. Barrett, C., Channakeshava, K., Huang, F., Kim, J., Marathe, A., Marathe, M.V., Pei, G., Saha, S., Subbiah, B. S. P., & Vullikanti, A. K. S. (2012). Human initiated cascading failures in societal infrastructures. *PLoS ONE, 7*(10), 1–20. http://dx.doi.org/10.1371%2Fjournal.pone.0045406
6. Bashan, A., Berezin, Y., Buldyrev, S. V., & Havlin, S. (2013). The extreme vulnerability of interdependent spatially embedded networks. *Nature Physics, 9*(10), 667–672. http://dx.doi.org/10.1038/nphys2727
7. Bassett, D. S., Porter, M. A., Wymbs, N. F., Grafton, S. T., Carlson, J. M., & Mucha, P. J. (2013). Robust detection of dynamic community structure in networks. *Chaos: An Interdisciplinary Journal of Nonlinear Science, 23*(1), 013142.
8. Battiston, F., Nicosia, V., & Latora, V. (2014). Structural measures for multiplex networks. *Physical Review E, 89*(3), 032804.
9. Baxter, G., Dorogovtsev, S., Goltsev, A., & Mendes, J. (2012). Avalanche collapse of interdependent networks. *Physical Review Letters, 109*(24), 248701.
10. Berlingerio, M., Coscia, M., Giannotti, F., Monreale, A., & Pedreschi, D. (2011). The pursuit of hubbiness: Analysis of hubs in large multidimensional networks. *Journal of Computational Science, 2*(3), 223–237.

11. Berlingerio, M., Coscia, M., Giannotti, F., Monreale, A., & Pedreschi, D. (2013). Multidimensional networks: Foundations of structural analysis. *World Wide Web, 16*(5–6), 567–593
12. Berlingerio, M., Pinelli, F., & Calabrese, F. (2013) Abacus: Frequent pattern mining-based community discovery in multidimensional networks. *Data Mining and Knowledge Discovery, 27*(3), 294–320.
13. Bianconi, G. (2013). Statistical mechanics of multiplex networks: Entropy and overlap. *Physical Review E, 87*(6), 062806.
14. Bródka, P., Kazienko, P., Musiał, K., & Skibicki, K. (2012). Analysis of neighbourhoods in multi-layered dynamic social networks. *International Journal of Computational Intelligence Systems, 5*(3), 582–596.
15. Bródka, P., Musial, K., & Kazienko, P. (2010). A method for group extraction in complex social networks. In *Knowledge management, information systems, E-learning, and sustainability research* (pp. 238–247).
16. Brodka, P., Stawiak, P., & Kazienko, P. (2011). Shortest path discovery in the multi-layered social network. In *2011 International Conference on Advances in Social Networks Analysis and Mining (ASONAM)* (pp. 497–501). Piscataway: IEEE.
17. Brummitt, C. D., Lee, K.-M., & Goh, K.-I. (2012). Multiplexity-facilitated cascades in networks. *Physical Review E, 85*, 045102. http://link.aps.org/doi/10.1103/PhysRevE.85.045102
18. Buldyrev, S. V., Parshani, R., Paul, G., Stanley, H. E., & Havlin, S. (2010). Catastrophic cascade of failures in interdependent networks. *Nature, 464*(7291), 1025–1028.
19. Cai, D., Shao, Z., He, X., Yan, X., & Han, J. (2005). Community mining from multi-relational networks. In *European Conference on Principles of Data Mining and Knowledge Discovery* (pp. 445–452). Berlin: Springer.
20. Carchiolo, V., Longheu, A., Malgeri, M., & Mangioni, G. (2011). Communities unfolding in multislice networks. In *Complex Networks* (pp. 187–195). Berlin: Springer.
21. Cardillo, A., Zanin, M., Gómez-Gardenes, J., Romance, M., del Amo, A. J. G., & Boccaletti, S. (2012). Modeling the multi-layer nature of the european air transport network: Resilience and passengers re-scheduling under random failures. arXiv:1211.6839.
22. Carley, K. M., Diesner, J., Reminga, J., & Tsvetovat, M. (2007). Toward an interoperable dynamic network analysis toolkit. *Decision Support Systems, 43*(4), 1324–1347.
23. Carley, K. M., & Hill, V. (2001). Structural change and learning within organizations. In *Dynamics of organizations: Computational modeling and organizational theories* (pp. 63–92).
24. Cascetta, E. (2009). *Transportation systems analysis: Models and applications* (vol. 29). New York: Springer Science & Business Media.
25. Cellai, D., López, E., Zhou, J., Gleeson, J. P., & Bianconi, G. (2013). Percolation in multiplex networks with overlap. *Physical Review E, 88*, 052811. http://link.aps.org/doi/10.1103/PhysRevE.88.052811
26. Chanda, S. (2013). *Petroleum pipelines: A handbook for onshore oil and gas pipelines*. Cambridge: Cambridge University Press.
27. Coscia, M., Rossetti, G., Pennacchioli, D., Ceccarelli, D., & Giannotti, F. (2013). "you know because i know": A multidimensional network approach to human resources problem. In *2013 IEEE/ACM International Conference on Advances in Social Networks Analysis and Mining (ASONAM)* (pp. 434–441). Piscataway: IEEE.
28. Cozzo, E., Arenas, A., & Moreno, Y. (2012). Stability of boolean multilevel networks. *Physical Review E, 86*, 036115. http://link.aps.org/doi/10.1103/PhysRevE.86.036115
29. Cozzo, E., Banos, R. A., Meloni, S., & Moreno, Y. (2013). Contact-based social contagion in multiplex networks. *Physical Review E, 88*(5), 050801.
30. Criado, R., Flores, J., García del Amo, A., Gómez-Gardeñes, J., & Romance, M. (2012). A mathematical model for networks with structures in the mesoscale. *International Journal of Computer Mathematics, 89*(3), 291–309.
31. Davis, D., Lichtenwalter, R., & Chawla, N. V. (2011). Multi-relational link prediction in heterogeneous information networks. In *2011 International conference on advances in social networks analysis and mining (ASONAM)* (pp. 281–288). Piscataway: IEEE.

32. D'Agostino, G., & Scala, A. (2014). *Networks of networks: The last frontier of complexity* (vol. 340). Berlin: Springer.
33. De Domenico M., Solé-Ribalta, A., Cozzo, E., Kivelä, M., Moreno, Y., Porter, M. A., Gómez, S., et al. (2013). Mathematical formulation of multilayer networks. *Physical Review X, 3*(4), 041022.
34. De Domenico, M., Solé-Ribalta, A., Gómez, S., & Arenas, A. (2014). Navigability of interconnected networks under random failures. *Proceedings of the National Academy of Sciences, 111*(23), 8351–8356.
35. Dickison, M., Havlin, S., & Stanley, H. E. (2012). Epidemics on interconnected networks. *Physical Review E, 85*(6), 066109.
36. Donges, J. F., Schultz, H. C., Marwan, N., Zou, Y., & Kurths, J. (2011). Investigating the topology of interacting networks. *The European Physical Journal B, 84*(4), 635–651.
37. Farid, A. M. (2015). Static resilience of large flexible engineering systems: Axiomatic design model and measures. *IEEE Systems Journal*. http://dx.doi.org/10.1109/JSYST.2015.2428284
38. Farid, A. M. (2016). A hybrid dynamic system model for multi-modal transportation electrification. *IEEE Transactions on Control System Technology*. http://dx.doi.org/10.1109/TCST.2016.2579602
39. Funk, S., & Jansen,V. A. (2010). Interacting epidemics on overlay networks. *Physical Review E, 81*(3), 036118.
40. Gan, L., Topcu, U., & Low, S. (2013). Optimal decentralized protocol for electric vehicle charging. *IEEE Transactions on Power Systems, 28*(2), 940–951.
41. Gao, L., Yang, J., Zhang, H., Zhang, B., & Qin, D. (2011). Flowinfra: A fault-resilient scalable infrastructure for network-wide flow level measurement. In *2011 13th Asia-Pacific network operations and management symposium*, p. KICS KNOM; IEICE ICM.
42. Gomez-Exposito, A., Conejo, A. J., & Canizares, C. (2008). *Electric energy systems: Analysis and operation*. Boca Raton, FL: CRC Press.
43. Gómez-Gardeñes, J., Reinares, I., Arenas, A., & Floría, L. M. (2012). Evolution of cooperation in multiplex networks. *Scientific Reports, 2*, 620.
44. Gong, Q., Midlam-Mohler, S., Serra, E., Marano, V., & Rizzoni, G. (2013). PEV charging control for a parking lot based on queuing theory. In *2013 American control conference* (pp. 1126–1131). Washington, DC: IEEE.
45. Harrer, A., & Schmidt, A. (2012). An approach for the blockmodeling in multi-relational networks. In *2012 IEEE/ACM International Conference on Advances in Social Networks Analysis and Mining (ASONAM)* (pp. 591–598). Piscataway: IEEE.
46. Havlin, S., Kenett, D., Bashan, A., Gao, J., & Stanley, H. (2014). Vulnerability of network of networks. *The European Physical Journal Special Topics, 223*(11), 2087–2106.
47. Hindes, J., Singh, S., Myers, C. R., & Schneider, D. J. (2013). Epidemic fronts in complex networks with metapopulation structure. *Physical Review E, 88*(1), 012809.
48. Horvát, E.-A., & Zweig, K. A. (2012). One-mode projection of multiplex bipartite graphs. In *Proceedings of the 2012 international conference on advances in social networks analysis and mining (ASONAM 2012)* (pp. 599–606). Washington, DC: IEEE Computer Society.
49. Irving, D., & Sorrentino, F. (2012). Synchronization of dynamical hypernetworks: Dimensionality reduction through simultaneous block-diagonalization of matrices. *Physical Review E, 86*(5), 056102.
50. Janić, M. (2014). *Advanced Transport Systems*. Berlin: Springer.
51. Kazienko, P., Musial, K., & Kajdanowicz, T. (2011). Multidimensional social network in the social recommender system. *IEEE Transactions on Systems, Man, and Cybernetics-Part A: Systems and Humans, 41*(4), 746–759.
52. Kazienko, P., Musial, K., Kukla, E., Kajdanowicz, T., & Bródka, P. (2011). Multidimensional social network: Model and analysis. *Computational Collective Intelligence. Technologies and Applications, 6922*, 378–387.
53. Kitamura, R., Kuwahara, M., & Kuwahara, M. (2005). *Simulation approaches in transportation analysis recent advances and challenges*. Tokyo, Japan: Springer.

54. Kivelä, M., Arenas, A., Barthelemy, M., Gleeson, J. P., Moreno, Y., & Porter, M. A. (2014) Multilayer networks. *Journal of Complex Networks, 2*(3), 203–271.
55. Lazega, E., Jourda, M.-T., Mounier, L., & Stofer, R. (2008). Catching up with big fish in the big pond? Multi-level network analysis through linked design. *Social Networks, 30*(2), 159–176.
56. Lee, K.-M., Kim, J. Y., Cho, W.-K., Goh, K.-I., & Kim, I. (2012). Correlated multiplexity and connectivity of multiplex random networks. *New Journal of Physics, 14*(3), 033027.
57. Leicht, E. A., & D'Souza, R. M. (2009). Percolation on interacting networks. ArXiv e-prints.
58. Lewis, T. G. (2011). *Network science: Theory and applications*. Hoboken, NJ: Wiley. http://books.google.ae/books?id=eVddjxBhLsoC
59. Li, W., Bashan, A., Buldyrev, S. V., Stanley, H. E., & Havlin, S. (2012). Cascading failures in interdependent lattice networks: The critical role of the length of dependency links. *Physical Review Letters, 108*, 228702. http://link.aps.org/doi/10.1103/PhysRevLett.108.228702
60. Louzada, V., Araújo, N., Andrade J. Jr., & Herrmann, H. (2013). Breathing synchronization in interconnected networks. arXiv:1304.5177.
61. Lurie, M. V. (2009). *Modeling of oil product and gas pipeline transportation*. Weinheim: Wiley-VCH Verlag GmbH & Co. KGaA.
62. Marceau, V., Noël, P.-A., Hébert-Dufresne, L., Allard, A., & Dubé, L. J. (2011). Modeling the dynamical interaction between epidemics on overlay networks. *Physical Review E, 84*(2), 026105.
63. Martin-Hernandez, J., Wang, H., Van Mieghem, P., & D'Agostino, G. (2013). On synchronization of interdependent networks. arXiv:1304.4731.
64. Menon, E. S. (2005). *Gas pipeline hydraulics*. Boca Raton, FL: CRC Press.
65. Min, B., Do Yi, S., Lee, K.-M., & Goh, K.-I. (2014). Network robustness of multiplex networks with interlayer degree correlations. *Physical Review E, 89*(4), 042811.
66. Min, B., & Goh, K. (2013). Layer-crossing overhead and information spreading in multiplex social networks. *Seed, 21*(T22), T12.
67. Möller, D. P. (2014). *Introduction to transportation analysis, modeling and simulation: Computational foundations and multimodal applications*. Berlin: Springer.
68. Mucha, P.J., Richardson, T., Macon, K., Porter, M.A., & Onnela, J.-P. (2010). Community structure in time-dependent, multiscale, and multiplex networks. *Science, 328*(5980), 876–878.
69. Newman, M. (2009). *Networks: An introduction*. Oxford, UK: Oxford University Press. http://books.google.ae/books?id=LrFaU4XCsUoC.
70. Ng, M. K.-P., Li, X., & Ye, Y. (2011). Multirank: co-ranking for objects and relations in multi-relational data. In *Proceedings of the 17th ACM SIGKDD international conference on Knowledge discovery and data mining* (pp. 1217–1225). New York: ACM.
71. Nicosia, V., Bianconi, G., Latora, V., & Barthelemy, M. (2013). Growing multiplex networks. *Physical Review Letters, 111*(5), 058701.
72. Parshani, R., Buldyrev, S. V., & Havlin, S. (2010). Interdependent networks: Reducing the coupling strength leads to a change from a first to second order percolation transition. *Physical Review Letters, 105*, 048701. http://link.aps.org/doi/10.1103/PhysRevLett.105.048701
73. Pattison, P., & Wasserman, S. (1999). Logit models and logistic regressions for social networks: Ii. multivariate relations. *British Journal of Mathematical and Statistical Psychology, 52*(2), 169–193.
74. Rocklin, M., & Pinar, A. (2013). On clustering on graphs with multiple edge types. *Internet Mathematics, 9*(1), 82–112.
75. Sahneh, F. D., Scoglio, C., & Chowdhury, F. N. (2013). Effect of coupling on the epidemic threshold in interconnected complex networks: A spectral analysis. In *American Control Conference (ACC), 2013* (pp. 2307–2312). Piscataway: IEEE.
76. Saumell-Mendiola, A., Serrano, M. A., & Boguñá, M. (2012). Epidemic spreading on interconnected networks. *Physical Review E, 86*, 026106. http://link.aps.org/doi/10.1103/PhysRevE.86.026106
77. Solá, L., Romance, M., Criado, R., Flores, J., García del Amo, A., & Boccaletti, S. (2013). Eigenvector centrality of nodes in multiplex networks. *Chaos: An Interdisciplinary Journal of Nonlinear Science, 23*(3), 033131.

78. Sole-Ribalta, A., De Domenico, M., Kouvaris, N. E., Díaz-Guilera, A., Gómez, S., & Arenas, A. (2013). Spectral properties of the laplacian of multiplex networks. *Physical Review E, 88*(3), 032807.
79. Sorrentino, F. (2012). Synchronization of hypernetworks of coupled dynamical systems. *New Journal of Physics, 14*(3), 033035.
80. Stroele, V., Oliveira, J., Zimbrao, G., & Souza, J. M. (2009). Mining and analyzing multirelational social networks. In *International conference on computational science and engineering, 2009. CSE'09* (vol. 4, pp. 711–716) Piscataway: IEEE.
81. Sun, W.-Q., Wang, C.-M., Song, P., & Zhang, Y. (2013). Flexible load shedding strategy considering real-time dynamic thermal line rating. *IET Generation, Transmission & Distribution, 7*(2), pp. 130–137
82. Sun, Y. (2012). Mining heterogeneous information networks. Ph.D. dissertation, University of Illinois at Urbana-Champaign.
83. Sun, Y., Han, J., Yan, X., Yu, P. S., & Wu, T. (2011). Pathsim: Meta path-based top-k similarity search in heterogeneous information networks. *Proceedings of the VLDB Endowment, 4*(11): 992–1003.
84. Sun, Y., Yu, Y., & Han, J. (2009). Ranking-based clustering of heterogeneous information networks with star network schema. In *Proceedings of the 15th ACM SIGKDD international conference on Knowledge discovery and data mining* (pp. 797–806). New York: ACM.
85. Tang, L., Wang, X., & Liu, H. (2012). Community detection via heterogeneous interaction analysis. *Data Mining and Knowledge Discovery, 25*(1), 1–33.
86. Tsvetovat, M., Reminga, J., & Carley, K. M. (2004). Dynetml: interchange format for rich social network data. *SSRN*. http://dx.doi.org/10.2139/ssrn.2729286
87. van Steen, M. (2012). *Graph theory and complex networks: An introduction*. Maarten van Steen.
88. Vazquez, A. (2006). Spreading dynamics on heterogeneous populations: Multitype network approach. *Physical Review E, 74*(6), 066114.
89. Wang, C., Lu, Z., & Qiao, Y. (2013). A consideration of the wind power benefits in day-ahead scheduling of wind-coal intensive power systems. *IEEE Transactions on Power Systems, 28*(1), 236–245.
90. Wasserman, S., & Faust, K. (1994). *Social network analysis: Methods and applications* (vol. 8). Cambridge: Cambridge University Press.
91. Wei, X., Valler, N., Prakash, B. A., Neamtiu, I., Faloutsos, M., & Faloutsos, C. (2012). Competing memes propagation on networks: A case study of composite networks. *ACM SIGCOMM Computer Communication Review, 42*(5), 5–12.
92. Xu, Y., & Liu, W. (2011). Novel multiagent based load restoration algorithm for microgrids. *IEEE Transactions on Smart Grid, 2*(1), 152–161.
93. Yağan, O., & Gligor, V. (2012). Analysis of complex contagions in random multiplex networks. *Physical Review E, 86*, 036103. http://link.aps.org/doi/10.1103/PhysRevE.86.036103
94. Yagan, O., Qian, D., Zhang, J., & Cochran, D. (2013). Conjoining speeds up information diffusion in overlaying social-physical networks. *IEEE Journal on Selected Areas in Communications, 31*(6), 1038–1048.
95. Zhang, H., Moura, S., Hu, Z., Qi, W., & Song, Y. (2017). Joint PEV charging station and distributed pv generation planning. In *2017 IEEE power & energy society general meeting* (pp. 1–5). Washington, DC: IEEE.
96. Zhou, D., Gao, J., Stanley, H. E., & Havlin, S. (2013). Percolation of partially interdependent scale-free networks. *Physical Review E, 87*(5), 052812.
97. Zhou, J., Xiang, L., & Liu, Z. (2007). Global synchronization in general complex delayed dynamical networks and its applications. *Physica A: Statistical Mechanics and its Applications, 385*(2), 729–742. http://linkinghub.elsevier.com/retrieve/pii/S0378437107007637

Chapter 3
Hetero-functional Graph Theory Preliminaries

This chapter is devoted to providing the reader with the fundamental concepts that are required for a solid understanding of the hetero-functional graph theory presented in Chapter 4. This foundation draws upon the theory of ontologies in Section 3.1. Section 3.2 then draws from the fields of systems engineering and engineering systems to present several definitions and concepts upon which hetero-functional graph theory is based.

3.1 Ontological Foundation for Hetero-functional Graph Theory

The modeling constraints found in multi-layer networks can be viewed from an ontological perspective. In the ontological sciences, the relationship between reality, the understanding of reality, and the description of reality is described by Ullman's Triangle [15]. Figure 3.1 displays Ullman's Triangle on the left, where reality is the *Real Domain* $\mathcal{D}$, understanding of reality is *Domain Conceptualization* $\mathcal{C}$, and the description of reality is *Language* $\mathcal{L}$. These general concepts are instantiated to describe modeling of real systems, where the reality is the *Physical System*, the modeler's understanding (or mental conceptualization) is the *Abstraction* $\mathcal{A}$, and the description of the abstraction is the *Model* $\mathcal{M}$. Figure 3.1 presents the instantiation of Ullman's general concepts on the right.

The modeling process abstracts the physical system to an abstraction, and represents the abstraction with a model [16]. The model refers to the physical system, but this reference is always indirect, as an abstraction is always made in the modeling process. The abstraction of reality may be entirely conceptual (residing within the mind) or linguistic (residing within some predefined language). In order for a model $\mathcal{M}$ to truly represent the abstraction $\mathcal{A}$, the modeling primitives of

W. C. H. Schoonenberg et al., *A Hetero-functional Graph Theory for Modeling Interdependent Smart City Infrastructure*,
https://doi.org/10.1007/978-3-319-99301-0_3

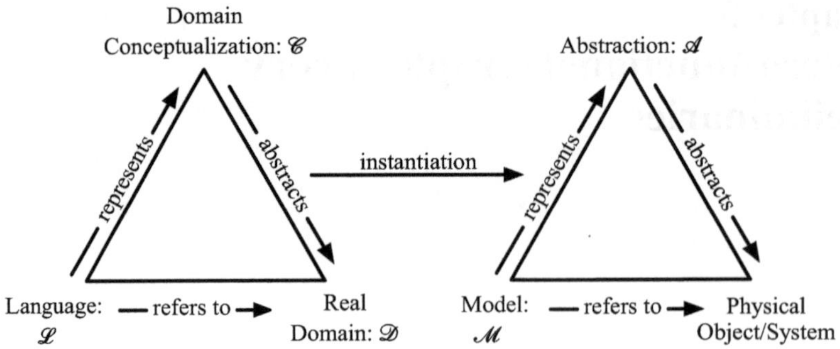

Fig. 3.1 Ullman's Triangle [16]: Its ontological definition. On the left, the relationship between reality, the understanding of reality, and the description of reality. On the right, the instantiated version of the definition

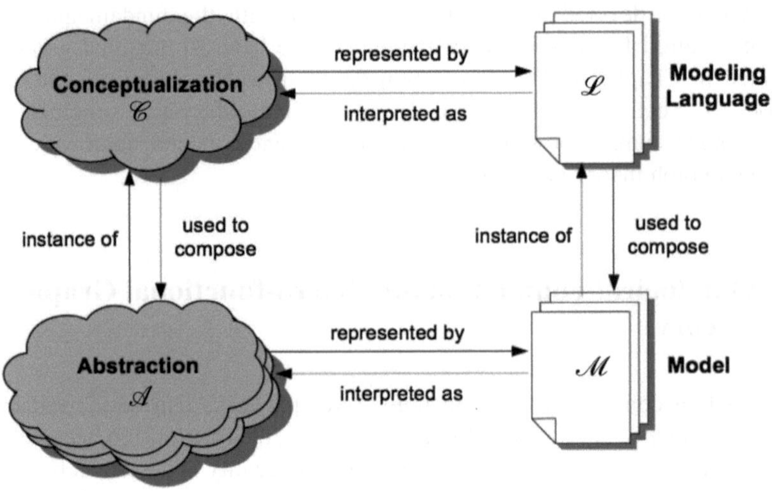

Fig. 3.2 The relationship between four ontological science concepts [16]: conceptualization, abstraction, modeling language, and model

the language $\mathcal{L}$ should "faithfully represent the domain conceptualization $\mathcal{C}$ to articulate the represented abstraction $\mathcal{A}$" [16]. In this definition, modeling primitives directly express relevant domain concepts, creating the domain conceptualization [16]. Visually, the relation between the Conceptualization, the Modeling Language, the Model, and the Abstraction is presented in Figure 3.2.

In this work, the abstraction is the class of interdependent infrastructure systems. The model, within the scope of this work, is assumed to be mathematical in nature, and more specifically, graph theoretic. The fidelity of the model with respect to an abstraction is determined by the four complementary linguistic properties shown in Figure 3.3 [16]: soundness, completeness, lucidity, and laconicity [15]. When all four properties are met, the abstraction and the model have an isomorphic (one-to-

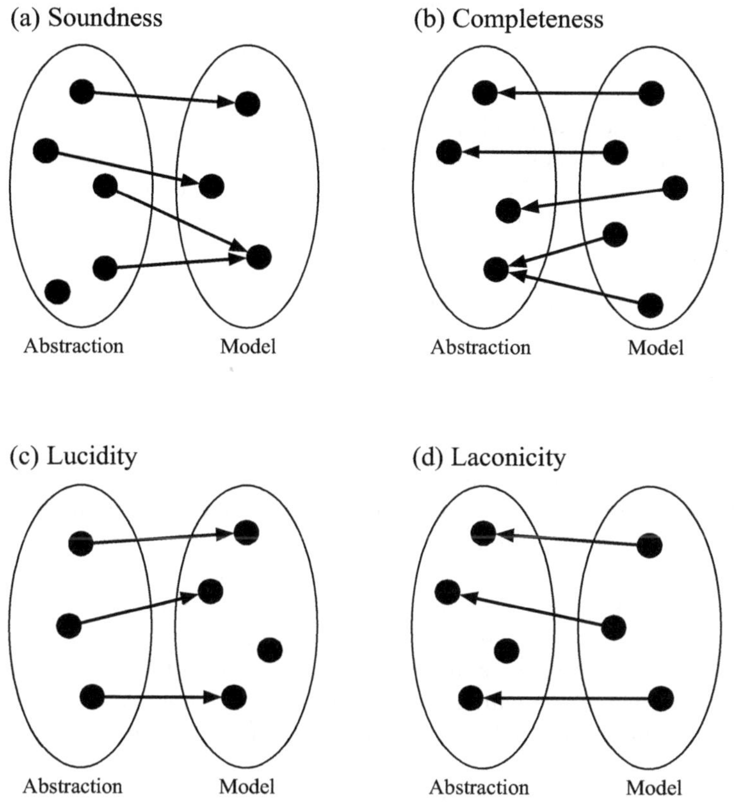

Fig. 3.3 Graphical representation of four ontological properties as mapping between abstraction and model: (**a**) Soundness, (**b**) Completeness, (**c**) Lucidity, and (**d**) Laconicity [16]

one) mapping and faithfully represent each other. The four properties are defined in general as Soundness, Completeness, Lucidity, and Laconicity:

Definition 3.1 – Soundness [15] A language $\mathcal{L}$ is sound w.r.t. a domain conceptualization $\mathcal{C}$ iff every modeling primitive in the language ($\mathcal{M}$) has an interpretation in the domain abstraction $\mathcal{A}$. (The absence of soundness results in the excess of modeling primitives w.r.t. the domain abstractions as shown in Figure 3.3.c on lucidity.) ∎

Definition 3.2 – Completeness [15] A language $\mathcal{L}$ is complete w.r.t. a domain conceptualization $\mathcal{C}$ iff every concept in the domain abstraction $\mathcal{A}$ of that domain is represented in a modeling primitive of that language. (The absence of completeness results in one or more concepts in the domain abstraction not being represented by a modeling primitive, as shown in Figure 3.3.d on laconicity.) ∎

Definition 3.3 – Lucidity [15] A language $\mathcal{L}$ is lucid w.r.t. a domain conceptualization $\mathcal{C}$ iff every modeling primitive in the language represents at most one domain concept in abstraction $\mathcal{A}$. (The absence of lucidity results in the overload of a modeling primitive w.r.t. two or more domain concepts as shown in Figure 3.3.a on soundness.) ■

Definition 3.4 – Laconicity [15] A language $\mathcal{L}$ is laconic w.r.t. a domain conceptualization $\mathcal{C}$ iff every concept in the abstraction $\mathcal{A}$ of that domain is represented at most once in the model of that language. (The absence of laconicity results in the redundancy of modeling primitives w.r.t the domain abstractions as shown in Figure 3.3.b on completeness.) ■

Given the discussion provided in Chapter 2, currently defined multi-layer networks maintain soundness and laconicity but lack completeness and lucidity. Reconsider Figure 2.1 on page 15.

- *Soundness* is maintained. There are no excess modeling primitives than those required by the example.
- *Completeness* is not maintained. The set of mathematical modeling elements are insufficient to represent all conceptual elements in the abstraction. For example, and as the following sections will discuss at length, multi-layer networks do not introduce the concept of system function. Reconsider nodes v_2 as a parking lot and v_3 as a charging station. Not only are charging stations and parking lots fundamentally different types of facilities, but the former has a superset of the functionality of the latter. While it is possible to reflect the hybrid functionality of the charging station with a mixed yellow and green circle (as in the figure), nowhere is this graphical depiction represented in the mathematics of multi-layer networks.
- *Lucidity* is not maintained. Not all conceptual elements in the abstraction have unique representations in the mathematical model. The representations are overloaded. For example, consider nodes v_2 as a parking lot and v_3 as a charging station. The same mathematical element of a green node is used to represent two fundamentally different abstractions of reality.
- *Laconicity* is maintained. There are no redundant or synonymous modeling primitives.

This ontological analysis of multi-layer networks suggests that its underlying modeling has several fundamental shortcomings. Instead, a different modeling language with a richer ontological foundation is required.

In order to define a Language $\mathcal{L}$, the Domain Conceptualization $\mathcal{C}$ needs to be clearly defined. This chapter continues to define the fundamentals of such a Domain Conceptualization in Section 3.2. Afterwards, Chapter 4 defines a language called hetero-functional graph theory.

3.2 Systems Engineering Foundations

This book looks to the field of systems engineering [25] for its domain conceptualization. Because the modern systems engineering field developed methodologically from industrial origins in the aerospace, communications, and defense sectors, it had to address a tremendous heterogeneity of applications in an immediately practical way.

Definition 3.5 – Systems Engineering (SE [25]) An interdisciplinary approach and means to enable the realization of successful systems. It focuses on defining customer needs and required functionality early in the development cycle, documenting requirements, and then proceeding with design synthesis and system validation while considering the complete problem: operations, cost and schedule, performance, training and support, testing, manufacturing, and disposal. SE considers both the business and the technical needs of all customers with the goal of providing a quality product that meets the user needs. ■

Originally, the practice of systems engineering was document-centric so as to capture customer requirements and trace them through the development and delivery of a system. As the field developed, engineering models became the center of practice.

Definition 3.6 – Model-Based Systems Engineering (MBSE [25]) The formalized application of modeling to support system requirements, design analysis, verification, and validation activities beginning in the conceptual design phase and continuing throughout development and later life-cycle phases. ■

The integration of modeling into the practice of MBSE immediately faced the practical challenge that no single model or modeling language was sufficient for the development of systems across the entirety of its scope throughout all stages of its engineering development [3, 14, 31]. Consequently, different models addressing different aspects and scopes of the overall system often became inconsistent [7, 19]. For example, a space shuttle requires a thermal-fluidic propulsion model, a structural model, a control system model, and an electrical model. All of which are coupled and must be synchronized as developments are made to each one.

To resolve this seemingly intractable problem, the field of MBSE developed the Systems Modeling Language (SysML) [14, 31] as an abstracted graphical model with sufficient ontological breadth to integrate and synchronize more detailed domain-specific engineering models. Here, the focus was not to develop complex mathematical models that provide engineering insight but rather to provide systems engineers and project managers with a tool by which to quickly understand the overall structure and behavior of a system and its component modules so as to coordinate its engineering development in large and often multiple engineering organizations [14, 31].

Recently, many have sought to use MBSE beyond the scope of complex products (in the aerospace and defense sectors) to large scale engineering systems (that form integral parts of smart cities). Examples of these include the power grid, transportation systems, healthcare delivery systems, and the internet.

Definition 3.7 – Engineering System [6] 1.) A class of systems characterized by a high degree of technical complexity, social intricacy, and elaborate processes aimed at fulfilling important functions in society. 2.) The term engineering systems is also used to refer to the engineering discipline that designs, analyzes, verifies, and validates engineering systems. ■

From an architectural perspective, these engineering systems are classified as large and flexible.

Definition 3.8 – Large Flexible Engineering System (LFES [9, 29]) An engineering system with a large set of system processes that not only evolve over time, but also can be fulfilled by one or more resources.[1] ■

This fact is important because hetero-functional graph theory was developed for this broad class of systems. Furthermore, large flexible engineering systems, as a class of systems, is characterized by a *meta-architecture* which generalizes and abstracts features found in reference and instantiated system architectures. Figure 3.4 uses a SysML Block Diagram to show how a meta-architecture generalizes a reference architecture which in turn generalizes an instantiated architecture. Consequently, the insights provided by hetero-functional graph theory at the meta-architecture level have direct pertinence to domain specific reference architectures and their instantiations in real-world applications.

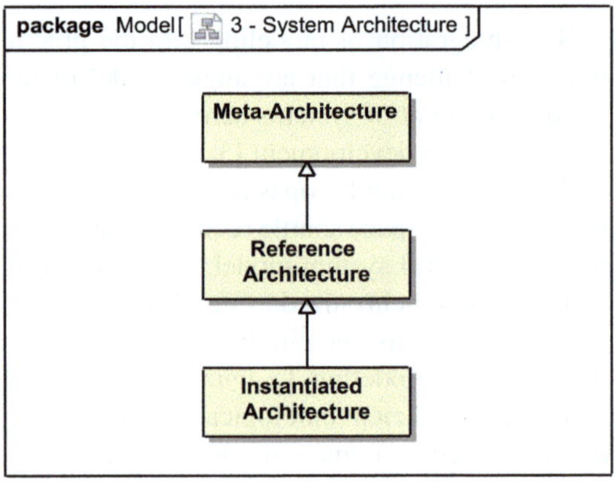

Fig. 3.4 SysML block diagram: system architecture can be represented at three levels of abstraction: instantiated, reference, and meta

[1]The original definition of large flexible engineering systems found in the Axiomatic Design literature [29] used the term functional requirements instead of system processes and the term design parameters instead of system resources. Recent works on hetero-functional graph theory [9, 11] have since adopted the definition revised as per the above for consistency of nomenclature.

For clarity, definitions for these terms are provided. A meta-architecture, a reference architecture, and an instantiated architecture are all types of (system) architectures.

Definition 3.9 – Systems Architecture [11] A system architecture consists of three parts, the physical architecture, the functional architecture, and their mapping. The *physical architecture* is a description of the partitioned elements of the system without any specification of the performance characteristics of the physical resources that comprise each element. The *functional architecture* is a description of the system processes in a solution-neutral way, structured in serial, or parallel, and potentially in hierarchical arrangements. The *system concept* as a mapping of the functional architecture onto the physical architecture completes the system architecture. ■

Instantiated architectures are the easiest to understand because they represent the wide variety of instantiated products and systems that everyone interacts with on a day-to-day basis. An instantiated systems architecture is now defined as:

Definition 3.10 – Instantiated Systems Architecture A case specific architecture, which represents a real-world scenario, or an example test case. At this level, the physical architecture consists of a set of instantiated resources, and the functional architecture consists of a set of instantiated system processes. The mapping defines which resources perform what processes. ■

Interestingly, in the case of large flexible engineering systems, the mapping of the instantiated functional architecture to the instantiated physical architecture is *necessarily* one-to-one [11]. Consequently, it adheres to the Independence Axiom in Axiomatic Design theory [12, 28, 29].

Axiom 3.1 – The Independence Axiom Maintain the independence of the system processes [12, 28, 29] such that: (1) they are mutually exclusive and collectively exhaustive with respect to each other, and (2) they maintain a one-to-one mapping with the system resources in the physical architecture.[2] ■

The adherence to the Independence Axiom is what allows a large engineering system to gain its flexible nature where paired elements of form and function can be routinely added or removed. Furthermore, the requirement for mutually exclusive and collectively exhaustive system processes effectively ensures the properties of completeness and laconicity.

Example 3.1 Consider the IEEE 201-bus power system test case in Figure 3.5 as an example of an instantiated system architecture. This test case has been used widely

---

[2]In certain cases the second condition can be relaxed so as to form a one-to-many mapping. However, the independence axiom explicitly prohibits many-to-many mappings of function to form.

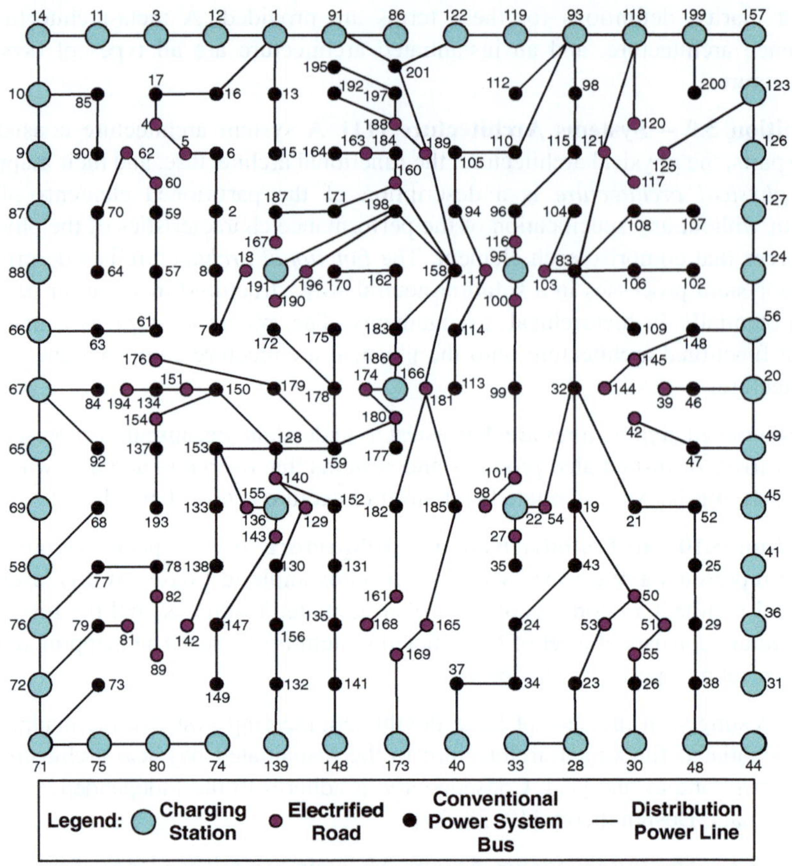

Fig. 3.5 201-Bus IEEE test case one line diagram [8, 10]

to perform power system performance and behavior studies [8, 10, 21, 24]. Based on Definitions 3.9 and 3.10, the instantiated system architecture of the test case has three parts:

- The *Physical Architecture* consists of 201 substation buses to which 1 generator and 95 loads are connected. There are also 200 branches that connect the buses. The physical lay-out of the test case is presented in Figure 3.5.
- The *Functional Architecture* is comprised of function *instances*. There is one function instance called "Generate Power." There are 95 function instances of "consume power"; each associated with the need to consume power at a given location. There are 201 function instances of "store power" as each substation must inevitably store some—albeit small—amounts power in order to route the electric power to other locations. Finally, there are 200 functions of "transport power" between two predetermined locations in space. Note that this description does not specify how these functions are fulfilled by engineering artifacts. The description is thus solution-neutral.

- The *System Concept* is the mapping of the functional architecture onto the physical architecture can be derived from the system specification. The 1 generator generates power, and the 95 loads consume power. The 201 substation buses store power, and the 200 branches transport power between their predefined locations in space.

In conclusion, the IEEE 201-bus power system test case represents an instantiated system architecture in that it provides a specific instance as a case study. It includes a unique layout and each element of the system architecture is named and countable.■

Reference architectures generalize instantiated system architectures. Instead of using individual instances as elements of its physical and functional architecture, it is expressed in terms of *domain-specific* classes of these instances. It is now defined as:

Definition 3.11 – Reference Architecture [5] *"The reference architecture captures the essence of existing architectures, and the vision of future needs and evolution to provide guidance to assist in developing new instantiated system architectures. …Such reference architecture facilitates a shared understanding across multiple products, organizations, or disciplines about the current architecture and the vision on the future direction. A reference architecture is based on concepts proven in practice. Most often preceding architectures are mined for these proven concepts. For architecture renovation and innovation validation and proof can be based on reference implementations and prototyping. In conclusion, the reference architecture generalizes instantiated system architectures to define an architecture that is generally applicable in a discipline. The reference architecture does however not generalize beyond its discipline."* ■

Example 3.2 Reconsider the IEEE 201-bus power system test case described in Example 3.1. The associated reference architecture describes power systems in general. Based on Definitions 3.9 and 3.11, the reference architecture of a power system has three parts:

- The *Physical Architecture* consists of buses, generators, loads, and branches without specifying either the number or names of their instances. The interfaces between these classes are also described in general. Generators and loads connect to individual buses[3] while branches connect to bus pairs. This reference physical architecture is graphically depicted with SysML in Figure 3.6. A common

[3]Power system test cases are often used to conduct power flow analysis studies. The underlying model neglects the lead lines from generators and loads to their associated buses. Other models of power systems such as transient stability models do include such lead lines. In such a case, the reference architecture would have no direct connection from the generators and loads to the buses. Furthermore, power flow analysis models usually do not differentiate between energy storage facilities and power generation facilities despite the former's ability to both generate and consume power.

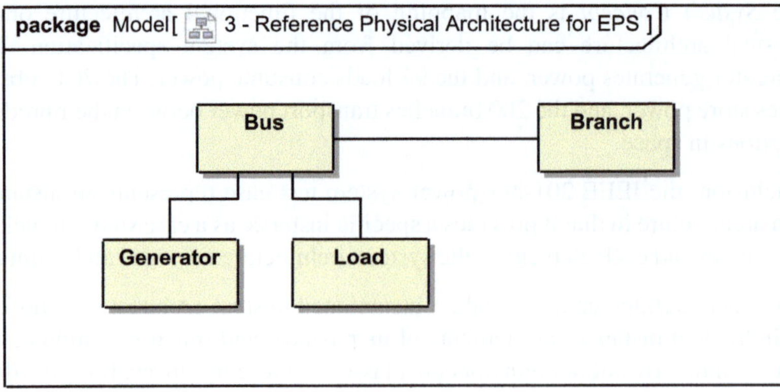

Fig. 3.6 Reference physical architecture for electric power systems

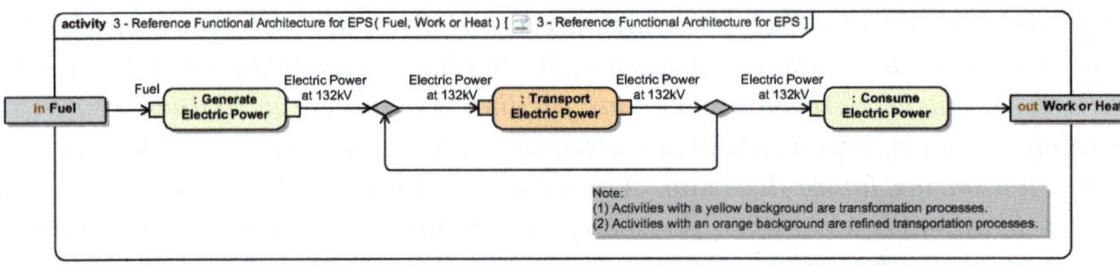

Fig. 3.7 Functional design pattern for an electric power system reference architecture

open-source power system solver uses this reference architecture to for its object-oriented software implementation [32, 33].

- The *Functional Architecture* comprises of three dominant functions (as types): Generate Power, Transport Power, and Consume Power. For mathematical simplicity and consistency with later sections, the storage of power is considered as transporting power between one place and itself. Again, this description of function does not describe how it is fulfilled and is therefore solution-neutral. This functional architecture is organized into a design pattern to display the system's behavior. Figure 3.7 shows the functional architecture of a power system as a design pattern. Note that the design pattern neither specifies the lay-out nor the number of elements in the power system, but only incorporates the functional arrangement of the system. The function of every power system, regardless of its physical architecture, is to generate power, transport it to the end users, who can then consume the power. The transportation of power may occur once, but usually occurs multiple times as there very rarely exists direct lines between power plants and end consumers.
- The *System Concept* maps the functional architecture onto the physical architecture. Given the functional and physical architectures at the instantiated and reference levels, and the system concept at the instantiated level, the system concept at the reference level can be straightforwardly deduced. The presence of

a mapping between a function instance to a physical elements causes a mapping between their associated function types and physical classes.

In all, the power system reference architectures presents a *domain-specific* representation of a power system's functional architecture, physical architecture, and system concept. ■

Meta-architectures further generalize reference architectures. Instead of domain-specific elements, it is expressed in terms of *domain-neutral* classes.

Definition 3.12 – Meta-Architecture A reference architecture composed of "primitive elements" that generalize the domain-specific functional and physical elements into their domain-neutral equivalents. ■

While no single engineering system architecture has been developed for all purposes, several modeling methodologies have been developed that span several discipline-specific domains. In the design of dynamic systems, bond graphs [2, 17, 23] and linear graphs [1, 4, 18, 20, 26] use generalized capacitors, resistors, inductors, gyrators, and transformers as primitive elements. In business dynamics, stocks and flows are often used as primitives [13, 27]. Finally, graph theory [22, 30] introduces nodes and edges as primitive elements. Each of these has its respective sets of applications. However, their sufficiency must ultimately be tested by an ontological analysis of soundness, completeness, lucidity, and laconicity.

Example 3.3 Reconsider the IEEE 201-bus power system test case described in Examples 3.1 and 3.2. The power system reference architecture can be described in a domain-neutral way. At the meta-level, the physical architecture becomes a graph with nodes that correspond to generators, loads, and buses and edges that correspond to power lines. The domain neutrality of a graph is evidenced by its application across multiple domains. A water distribution system, for example, can be presented as a graph where the nodes and edges correspond to water junctions and pipelines. That said, this graph is only a representation of the power system's physical architecture at the meta level. It represents neither the functional architecture nor the system concept. In that regard, it is an ontologically inadequate representation of the system's meta-architecture as a whole. The exposition of hetero-functional graph theory in the following chapter serves to address this issue of ontological sufficiency. ■

In conclusion, the field of systems engineering has evolved tremendously in recent decades; first with the incorporation of modeling tools like SysML to become MBSE, and then second with an expansion of scope to engineering systems to become the engineering systems discipline. In that regard, the application of MBSE tools such as SysML is indispensable. The systems can furthermore be described at several levels of abstraction: the meta-architecture, the reference architecture, and the instantiated system architecture levels, respectively. Graph Theory is an example of a meta-architecture, capable of describing domain-specific systems in a general manner. The theory presented in this book, called hetero-functional graph theory, presents itself to address the ontological insufficiencies of graphs defined purely in terms of nodes and edges. The book now continues with an exposition of hetero-functional graph theory in Chapter 4.

References

1. Blackwell, W. A. (1968). *Mathematical modeling of physical networks*. Collier-Macmillan, New York.
2. Brown, F. T. (2007). *Engineering system dynamics* (2nd ed.). Boca Raton, FL: CRC Press.
3. Buede, D. M. (2009). *The engineering design of systems: Models and methods* (2nd ed.). Hoboken, NJ: Wiley.
4. Chan, S.-P., Chan, S.-Y., & Chan, S.-G. (1972). *Analysis of linear networks and systems*. Reading. MA: Addison-Wesley.
5. Cloutier, R., Muller, G., Verma, D., Nilchiani, R., Hole, E., & Bone, M. (2010). The concept of reference architectures. *Systems Engineering, 13*(1), 14–27.
6. De Weck, O. L., Roos, D., & Magee, C. L. (2011). *Engineering systems: Meeting human needs in a complex technological world*. Cambridge, MA: MIT Press. http://www.knovel.com/knovel2/Toc.jsp?BookID=4611http://mitpress-ebooks.mit.edu/product/engineering-systems
7. Dori, D. (2015). *Model-based systems engineering with OPM and SysML*. Berlin: Springer.
8. Farid, A. M. (2015). Symmetrica: Test case for transportation electrification research. *Infrastructure Complexity, 2*(9), 1–10. http://dx.doi.org/10.1186/s40551-015-0012-9
9. Farid, A. M. (2015). Static resilience of large flexible engineering systems: Axiomatic design model and measures. *IEEE Systems Journal*. http://dx.doi.org/10.1109/JSYST.2015.2428284
10. Farid, A. M. (2016). A hybrid dynamic system model for multi-modal transportation electrification. *IEEE Transactions on Control System Technology*. http://dx.doi.org/10.1109/TCST.2016.2579602
11. Farid, A. M. (2016). An engineering systems introduction to axiomatic design. In Farid, A. M., & Suh, N. P. (Eds.), *Axiomatic design in large systems: Complex products, buildings & manufacturing systems* (ch. 1, pp. 1–47). Berlin: Springer. http://dx.doi.org/10.1007/978-3-319-32388-6
12. Farid, A. M., & Suh, N. P. (2016). *Axiomatic design in large systems: Complex products, buildings and manufacturing systems*. Berlin: Springer. http://dx.doi.org/10.1007/978-3-319-32388-6
13. Forrester, J. W. (1958). Industrial dynamics: A major breakthrough for decision makers. *Harvard Business Review, 36*(4), 37–66.
14. Friedenthal, S., Moore, A., & Steiner, R. (2011). *A practical guide to SysML: The systems modeling language* (2nd ed.). Burlington, MA: Morgan Kaufmann
15. Guizzardi, G. (2005). *Ontological foundations for structural conceptual models*. CTIT, Centre for Telematics and Information Technology.
16. Guizzardi, G. (2007). On ontology, ontologies, conceptualizations, modeling languages, and (meta) models. *Frontiers in Artificial Intelligence and Applications, 155*, 18.
17. Karnopp, D., Margolis, D. L., & Rosenberg, R. C. (1990). *System dynamics: A unified approach* (2nd ed). New York: Wiley. http://www.loc.gov/catdir/enhancements/fy0650/90012110-t.html
18. Koenig, H. E., Tokad, Y., & Kesavan, H. K. (1967). *Analysis of discrete physical systems*. McGraw-Hill, New York.
19. Kossiakoff, A., & Sweet, W. N. (2003). Knovel (Firm). In *Systems engineering principles and practice*. Hoboken, NJ: Wiley. http://www.knovel.com/knovel2/Toc.jsp?BookID=1430
20. Kuo, B. C. (1967). *Linear networks and systems*. New York: McGraw-Hill.
21. Manuel de Oliveira de Jesus, P. (2007). *Remuneration of distributed generation: A holistic approach*. Ph.D. dissertation, Faculdade de Engharia Universidade de Porto.

22. Newman, M. (2009). *Networks: An introduction*. Oxford, UK: Oxford University Press. http://books.google.ae/books?id=LrFaU4XCsUoC
23. Paynter, H. M. (1961). *Analysis and design of engineering systems*. Cambridge: MIT Press.
24. Ramirez-Rosado, I. J., & Bernal-Agustin, J. L. (1998). Genetic algorithms applied to the design of large power distribution systems. *IEEE Transactions on Power Systems, 13*(2), 696–703.
25. SE Handbook Working Group. (2015). *Systems engineering handbook: A guide for system life cycle processes and activities*. International Council on Systems Engineering (INCOSE).
26. Shearer, J. L., Murphy, A. T., & Richardson, H. H. (1967). *Introduction to system dynamics* (vol. 7017). Reading, MA: Addison-Wesley.
27. Sterman, J. D. (2000). *Business dynamics: Systems thinking and modeling for a complex world* (vol. 19). Boston, MA: Irwin/McGraw-Hill.
28. Suh, N. P. (1990). *The principles of design*. Oxford: Oxford University Press.
29. Suh, N. P. (2001). *Axiomatic design: Advances and applications*. Oxford: Oxford University Press.
30. van Steen, M. (2010). *Graph theory and complex networks: An introduction*. Maarten van Steen
31. Weilkiens, T. (2007). *Systems engineering with SysML/UML modeling, analysis, design*. Burlington, MA: Morgan Kaufmann.
32. Zimmerman, R. D., & Murillo-Sanchez, C. E. (2011). Matpower 4.1 user's manual. Power Systems Engineering Research Center, Tech. Rep.
33. Zimmerman, R. D., Murillo-Sanchez, C. E., & Thomas, R. J. (2011). MATPOWER: Steady-state operations, planning, and analysis tools for power systems research and education. *IEEE Transactions on Power Systems, 26*(1), 12–19. http://dx.doi.org/10.1109/TPWRS.2010.2051168

Chapter 4
Hetero-functional Graph Theory

As mentioned in the book's preface, hetero-functional graph theory can be viewed as an intellectual fusion of model-based systems engineering and network science. Nevertheless, the theory has origins elsewhere. More specifically, the original target application of hetero-functional graph theory was the automated mass-customized production system literature [36, 37, 43–45]. There, it was recognized that the need to compete in dynamic marketplaces with products of increasingly short product life-cycle drove mass-customized production systems to explicitly foster reconfigurability as a life-cycle property of their integrated automation solutions. The first publication on hetero-functional graph theory appeared in 2006 to address system degrees of freedom (as defined in Definition 4.13) [20]. One year later, *reconfigurability measurement in automated manufacturing systems* appeared in 2007 as the first comprehensive treatment of hetero-functional graph theory [8]. It utilized a design structure matrix as a type of graph to address the ease of reconfiguration [9, 21]. Furthermore, it drew the concept of a knowledge base from the Axiomatic Design literature [46] and quantified it as a bipartite graph that allocates function to form [10, 22]. Beyond these considerations, the reconfigurability measurement of mass-customized production systems needed to specifically address heterogeneity as its essential characteristic. Finally, because automation was an essential aspect of mass-customized production systems, the theory was explicitly cyber-physical. Since that time, and as will be discussed in greater detail in Chapter 6, hetero-functional graph theory has been applied in numerous domains with several enhancements to address the peculiarities of each application. As this chapter lays out hetero-functional graph theory, historical footnotes are provided to describe how the theory has developed over time.

This chapter provides an exposition of hetero-functional graph theory in terms of its constituent mathematical models and how they relate to their counterparts in SysML. These models are: (1) the System Concept, (2) the Hetero-functional Adjacency Matrix, (3) the Controller Agency Matrix, (4) the Controller Adjacency Matrix, (5) the Service as Operand Behavior, (6) the Service Feasibility Matrix,

W. C. H. Schoonenberg et al., *A Hetero-functional Graph Theory for Modeling Interdependent Smart City Infrastructure*,
https://doi.org/10.1007/978-3-319-99301-0_4

and (7) the System Adjacency matrix. The first two models are assumed to be universal and apply to all types of engineering systems. They form the structural model. The last four apply when it is necessary to differentiate systems. Models 3 and 4 constitute the system control model. Together, they differentiate systems based upon the structure of their control and decision-making. Models 5 and 6 constitute the service model. They differentiate systems based upon the behavior of their operands. These six models are then ultimately coupled together for holistic analysis of cyber-physical engineering systems. The structural model and the control model have cyber-physical resource interfaces. Similarly, the structural model and the service model have a structure-service coupling. When integrated together, these models constitute the final product of hetero-functional graph theory: the System Adjacency Matrix. Table 4.1 presents a visual overview of the models.

Example 4.1 In order to facilitate the theoretical discussion of hetero-functional graph theory, Figure 4.1 introduces a simplistic example network of a smart city. The network consists of four nodes and four edges. The nodes are: (1) water treatment facility, (2) solar PV system, (3) house, and (4) work location. The edges are: (1) water pipeline from the water treatment facility to the house, (2) electric power line from the solar PV system to the water treatment facility, (3) electric power line from the solar PV system to the house, and (4) road from the work location to the house, and back. ■

4.1 System Concept

Building upon the Definition 3.9 of System Architecture in Section 3.2 on page 29, a system's concept is the allocation of a system's *function* to its *form* [2]. The system function is described by elements that include verbs acting upon their operands. The system form is described with elements that include nouns. When an element of function is allocated to an element of form, a complete sentence is formed consisting of a noun-subject, its active transitive verb, and its operand. The sentence as a whole represents a physical *capability* in the system.

SysML consists of several graphical diagrams that represent an engineering system [25, 50]. Of these, the block diagram primarily represents system form while the activity diagram represents system function. Because there is no dedicated diagram for system concept, it appears in both diagrams. Consider the SysML block diagram in Figure 4.2. It shows a model of a smart city's form consisting of several types of resources (as form elements) which are capable of completing several processes (as function elements). This leads to sentence constructions like "Water pipeline transports water." Now consider the activity diagram with swim lanes in Figure 4.3. It shows a model of a smart city's function consisting of a logical flow of processes which have been allocated to several resources in vertical columns called "swim lanes." In each case, the system concept is represented equivalently.

Table 4.1 An overview of the mathematical models in hetero-functional graph theory: the shaded area maps mathematical elements to their associated models

Elements	Mathematical Elements	Structural Model	Control Model	Cyber-Physical Resource Interfaces	Service Model	Structure-Service Coupling	System Adjacency Matrix
I: System Concept ---- Chapter 4.1	System Function (Processes)	■				■	■
	System Form (Physical Resources)	■					■
	System Knowledge Base	■					■
	System Constraints Matrix	■					■
	Structural Degrees of Freedom	■				■	■
II: Hetero-functional Adjacency Matrix ---- Chapter 4.2	System-Sequence Knowledge Base	■		■			■
	System-Sequence Constraints Matrix	■		■			■
	Hetero-functional Adjacency Matrix	■		■			■
	System-Sequence Degrees of Freedom	■		■			■
III: Controller Agency Matrix ---- Chapter 4.3	Physical Resources		■	■			■
	(Cyber-)Resources		■	■			■
	Controller Agency Matrix		■	■			■
IV: Controller Adjacency Matrix ---- Chapter 4.4	Controller Adjacency Matrix		■	■			■
V: Service as Operand Behavior ---- Chapter 4.5	Services				■	■	■
	Service Activities				■	■	■
	Service String and Service Petri net				■	■	■
VI: Service Feasibility Matrix ---- Chapter 4.6	Service Transformation Feasibility Matrix				■	■	■
	Service Transportation Feasibility Matrix				■	■	■
	Service Line Feasibility Matrix				■	■	■
	Service Degrees of Freedom				■	■	■
VII: System Adjacency Matrix ---- Chapter 4.7	System Adjacency Matrix	■	■	■	■	■	■

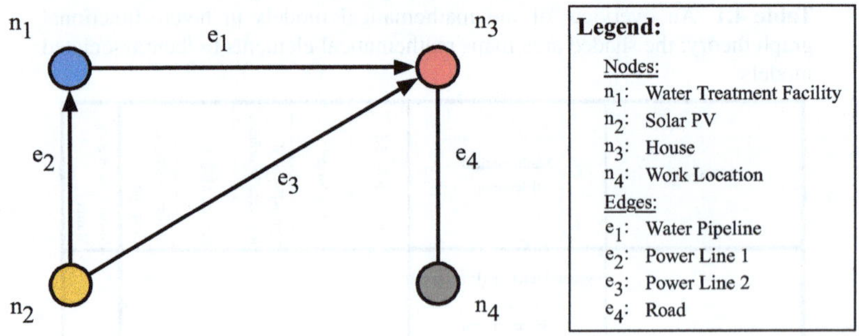

Fig. 4.1 An Example 4-Node Smart City Network: a simplistic smart city network that is used as an example throughout this chapter

Hetero-functional graph theory now represents this SysML description of function allocated to form mathematically [8, 10, 14, 17, 20, 22].

4.1.1 System Form

First, it defines a set of system resources R as form elements. These are then classified into three types. $R = M \cup B \cup H$, where M are the transformation resources (or machines), B are the independent buffers, and H are the transporting resources.[1]

Definition 4.1 – Transformation Resource: A resource $r \in R$ is a transformation resource $m \in M$ iff it is capable of one or more transformation processes on one or more operands and it exists at a unique location in space. ■

Definition 4.2 – Independent Buffer: A resource $r \in R$ is an independent buffer $b \in B$ iff it is capable of storing one or more operands and is not able to transform them or transport them to another location and it exists at a unique location in space. ■

Definition 4.3 – Buffer: A resource $r \in R$ is a buffer $b_s \in B_S$ iff it is capable of storing one or more operands at a unique location in space. $B_S = M \cup B$. ■

Definition 4.4 – Transportation Resource: A resource $r \in R$ is a transportation resource $h \in H$ iff it is capable of transporting one or more operands between an origin and a distinct destination, without transforming these operands. ■

[1]The resource definitions presented above were mildly revised from those found in [8, 10, 14, 17, 20, 22] so as to maintain the integrity of the ontological argument presented throughout the book.

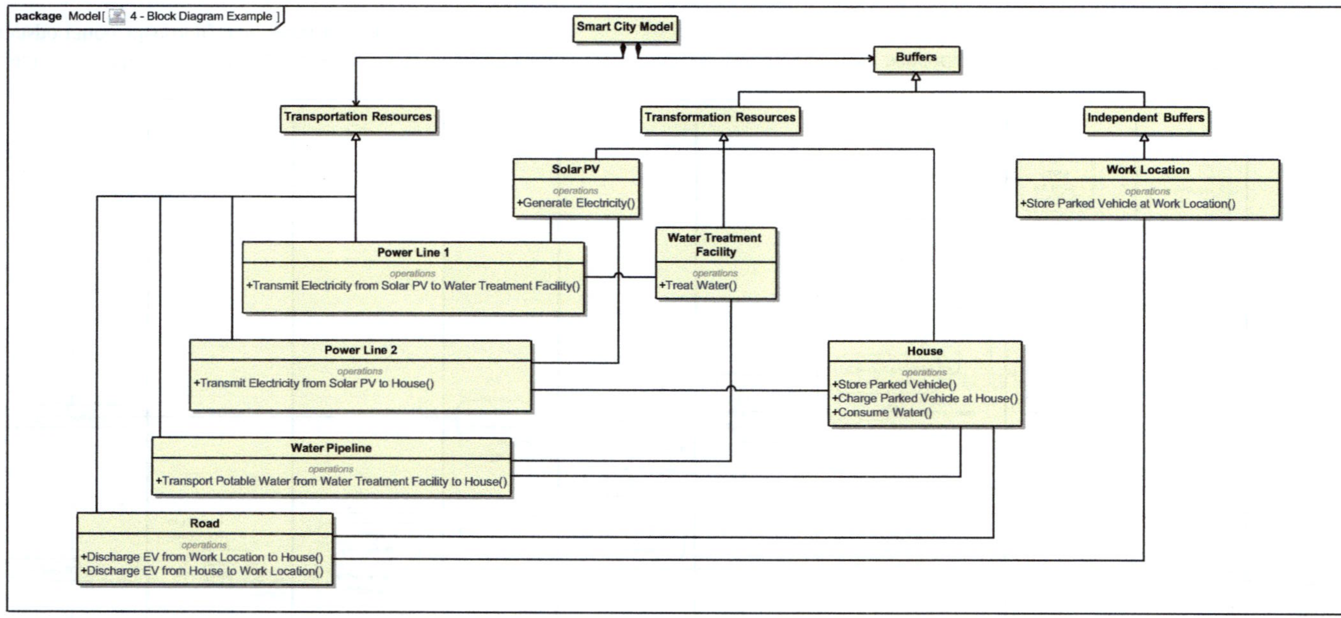

Fig. 4.2 A SysML Block Diagram: a representation of Figure 4.1 using the SysML. The 4-node smart city network consists of transportation, electricity, and water infrastructure

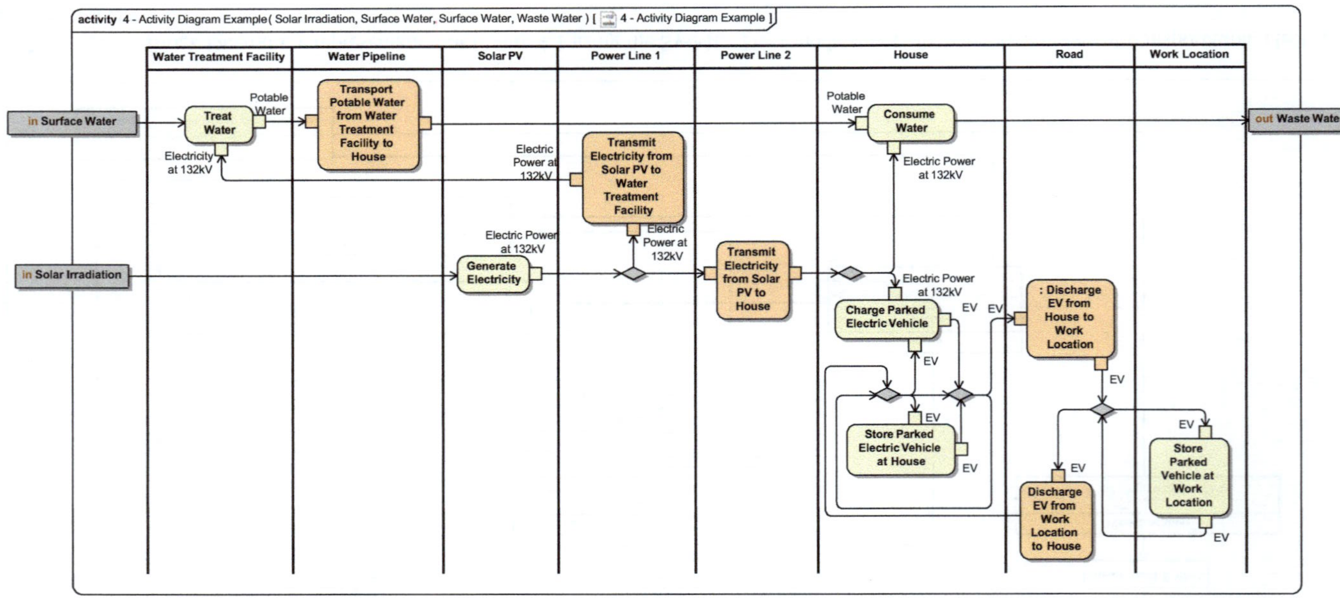

Fig. 4.3 A SysML Activity Diagram: swim lanes allocate function to form for the 4-node smart city network as presented in Figure 4.1. The network consists of transportation, electricity, and water infrastructure

Ontological Remark 4.1 *The classification of system resources into three different types originates from the field of production systems where each type of resource is conceptualized to have its own type of value. Transformation resources are often times referred to as "value adding," transportation resources gain value from logistical necessity, and independent buffers are often explicitly minimized. The distinction between these different types of resources serves to enhance the ontological* lucidity *of hetero-functional graph theory with respect to the conceptualization of value and avoid construct overload. Furthermore, the use of iff conditionals in the definition of these types of resources guarantees adherence to all four ontological properties.* ■

4.1.2 System Function

Next, hetero-functional graph theory defines a set of physical system processes P as functional elements.

Definition 4.5 – System Process [27, 41]: An activity that transforms a predefined set of inputs into a predefined set of outputs. ■

The system processes are then classified into two types. $P = P_\mu \cup P_\eta$, where P_μ are the transformation processes,[2] and P_η are the transportation processes between buffers $B_S = M \cup B$.[3]

Definition 4.6 – Transformation Process: A process is a transformation process $p_{\mu j} \in P_\mu$ iff it is capable of transforming one or more properties of a set of operands into a distinct set of output properties in place. It's syntax is:

$$\{\text{transitive verb, operands}\} \rightarrow \{\text{outputs}\} \tag{4.1}$$

■

Definition 4.7 – Transportation Process: A process is a transportation process $p_{\eta u} \in P_\eta$ iff it is capable of transporting one or more operands between an origin buffer $b_{sy_1} \in B_S$ to a destination buffer $b_{sy_2} \in B_S$ according to the following convention of indices [8, 10, 14, 20, 22][4]:

$$u = \sigma(B_S)(y_1 - 1) + y_2 \tag{4.2}$$

[2]The transformation processes are assumed to transform the operand in place.

[3]The system process definitions presented above were mildly revised from those found in [8, 10, 14, 17, 20, 22] so as to maintain the integrity of the ontological argument presented throughout the book.

[4]Note that a "storage process" is merely a transportation process with the same origin and destination.

It's syntax is:

$$\{\text{transport, operands,origin, destination}\} \rightarrow \{\text{outputs, destination}\} \tag{4.3}$$

■

Ontological Remark 4.2 *The classification of system processes into two different types originates from the field of production systems where each type of processes is conceptualized to have its own type of value. Transformation processes are often times referred to as "value adding," while transportation processes gain value from logistical necessity. The distinction between these different types of processes serves to enhance the ontological* lucidity *of hetero-functional graph theory with respect to the conceptualization of value and avoid construct overload. Furthermore, the use of iff conditionals in the definition of these types of processes guarantees adherence to all four ontological properties.* ■

Furthermore, the Independence Axiom (3.1) requires the mutual exclusivity of system processes.

Theorem 4.1 – Mutual Exclusivity of System Processes: A lucid representation of system processes as a domain conceptualization distinguishes between two system processes as modeling primitives with different sets of inputs and outputs.■

Proof 4.1 Assume that the inputs or outputs of a system process conceptualization are distinct from the inputs or outputs of another system process conceptualization. By Definition 4.5, these two system process conceptualizations are distinct. If a single system process modeling primitive is used to represent both of these two system process conceptualizations, then it is overloaded and by definition lucidity is violated. Consequently, two system process modeling primitives are required. ■

The above defined set of transportation processes, however, does not distinguish the multiple ways in which an operand is transported between origin and destination. Therefore, at times, it is necessary to introduce a set of *refined* transportation processes $P_{\bar{\eta}}$ which are defined individually as a feasible combination of a transportation process $p_\eta \in P_\eta$ and a holding process $p_\gamma \in P_\gamma$.

Definition 4.8 – Holding Process: A process is a holding process $p_{\gamma g} \in P_\gamma$ iff it holds one or more operands during the transportation from one buffer to another. ■

In order to maintain the independence axiom and the mutual exclusivity of the system processes (Theorem 4.1), holding processes are specified so as to distinguish between transportation processes that:

- Have *different* operands,
- Hold a given operand in a *given way*, or
- Change the *state* of the operand.

Note that the last requirement for holding processes give them a potentially transformative nature.[5] Furthermore, in differentiating transportation processes by holding process, it is important to retain one holding process that has no transformative effect on its operand.

The set of refined transportation processes can be expressed as a combination of the set of transportation processes and the set of holding processes:

$$P_{\bar{\eta}} = P_{\gamma} \bigtimes P_{\eta} \tag{4.4}$$

where $\bigtimes$ is the Cartesian Product [49].

Definition 4.9 – Refined Transportation Process: A process is a refined transportation process $p_{\bar{\eta}\varphi} \in P_{\bar{\eta}}$ iff it is capable of transporting one or more operands between an origin buffer $b_{sy_1} \in B_S$ to a destination buffer $b_{sy_2} \in B_S$ while it is realizing holding process $p_{\gamma g} \in P_{\gamma}$. Its syntax is:

$$\{\text{transport, operands, origin, destination, while transitive verb}\} \rightarrow \{\text{outputs, destination}\} \tag{4.5}$$

■

In systems where holding processes are required, the complete set of system processes becomes: $P = P_{\mu} \cup P_{\bar{\eta}}$.

The introduction of holding processes of a transformative nature does require special attention. The use of hetero-functional graph theory must now take care to not double count transformation processes and refined transportation processes with a transformative nature that occur in a single location. This work uses transformation processes (rather than holding processes of a transformative nature) by default. It only uses transformative holding processes under two conditions: (1) they are required with transportation processes with a distinct origin and destination, or (2) the value-related operand [4] of the process (e.g., pumped water in a water distribution system) remains in place while a second operand is transformed (e.g., electricity). For the sake of simplicity, the transformative nature of such a refined transportation process is modeled to occur at a single location in space. For example, while an electric pump may pump water from one distinct location to another, the transformative nature of consuming electricity occurs at a single point in the electric distribution system. Finally, resources that are capable of such refined transformation processes with a transformative nature would now be classified as transformation resources by virtue of Definition 4.1.

---

[5]While holding processes have been a part of hetero-functional graph theory since its inception, this work is the first to allow for the holding processes to change the state of the operand and have a potentially transformative nature. As later chapters will demonstrate, this expansion of the meaning of holding processes is necessitated by interdependent smart city infrastructure applications where a given operand changes its state as it moves from one location to another.

Table 4.2 System processes and resources

System processes		System resources	
P_μ	Transformation processes	M	Transformation resources
P_η	Transportation processes	H	Transportation resources
P_γ	Holding processes	B	Independent buffers
$P_{\bar{\eta}} = P_\gamma \times P_\eta$	Refined transportation processes	$B_S = M \cup B$	Buffers
$P = P_\mu \cup P_{\bar{\eta}}$	System processes	$R = M \cup B_S$	System resources

Based on the discussion above, one could derive that a separate set of transformation processes P_μ is no longer necessary as the holding processes can introduce transformative nature to a buffering transporting processes. However, transformation processes are *essential* to represent operands crossing the system boundary. When the operand enters the system, it seems that the operand appears from "nowhere." The operand is transported across the system boundary, and appears the moment it enters the system. The transportation process has an origin outside the system boundary, which consequently does not exist. Consequently, these processes are necessarily represented by transformation processes, as if these operands were generated or consumed at the point of entry or departure. The refined transportation processes of transformative nature convert the state of the transported operand, but the transported operand is always retained in the system. Heat losses or waste can be included as an output operand of the process,[6] but can only leave the system via a transformation process.

In keeping with the ontological properties described above, hetero-functional graph theory requires that the set of system resources and processes be *mutually exclusive and collectively exhaustive* descriptions of system form and function, respectively. The overview of system processes and resources is provided in Table 4.2.

Ontological Remark 4.3 *The above paragraphs imply an underlying meta-architecture. In hetero-functional graph theory, the primitives of the meta-architecture of system form is a set of resources; classified as transportation resources H and buffers B_S which are in turn classified as transformation resources M and independent buffers B. This classification is depicted in Figure 4.4. Meanwhile, the meta-architecture of system function is a set of processes; classified as transformation processes P_μ, transportation processes P_η, and holding processes P_γ. It is depicted in Figure 4.5 as a SysML activity diagram. Here, all potential*

[6]These outputs may also be neglected when they are not desired in the model.

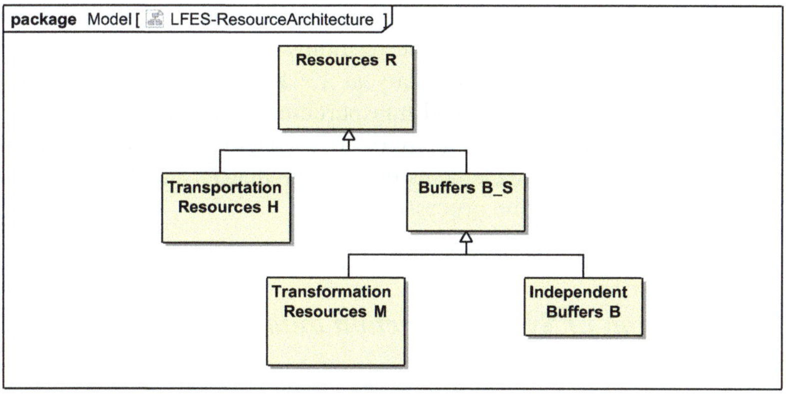

Fig. 4.4 A SysML Block Diagram: the meta-architecture of the system form of an LFES

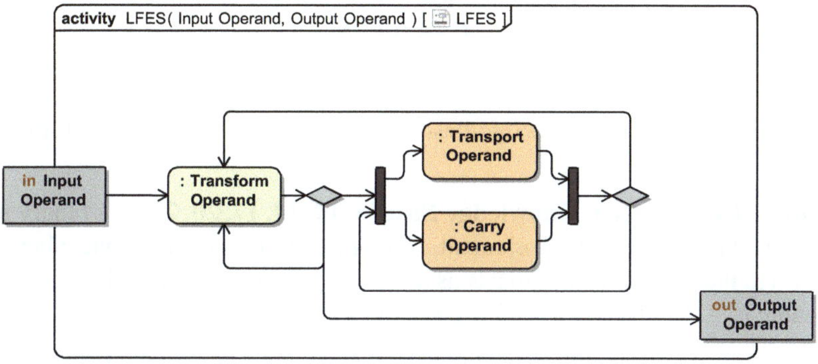

Fig. 4.5 A SysML Activity Diagram: the meta-architecture of the system function of an LFES

sequences of processes are indicated because there is no a priori reason at the meta-level to prevent one process from following another. ■

Example 4.2 The SysML block diagram in Figure 4.2 represents a simplified model of a smart city infrastructure system and is now modeled using hetero-functional graph theory. The system resources R are M={Water Treatment Facility, Solar PV, House}. B={Work Location}. H={Water Pipeline, Power Line, Road}. Now, reconsider the activity diagram with swim lanes in Figure 4.3. The activity diagram has 2 inputs, 1 output, and 11 activities allocated to 8 resources. First, the set of transformation processes is P_μ = {Treat water, consume water, generate electricity}. Note that all three transformation processes either have an input as operand or their output is also a system output. Second, the set of transportation processes P_η is defined as all transportation processes between the system buffers B_S. This example has four buffers ($B_S = M \cup B$) and consequently, the example has 16 transportation processes (e.g., transport from M_1 to M_1, transport from M_1 to M_2, etc.). Third, the set of holding processes serves to distinguish transportation processes. The transportation processes, in this example, are distinguished for operand and transformative nature. The resulting set of holding processes is P_γ = {Carry potable water, Carry electricity at 132 kV, Charge EV, Discharge EV, Carry EV}. Finally, the set of refined transportation processes is calculated using Equation 4.4. The resulting set has a size of 80, it contains all unique combinations of the elements in P_γ with the elements in P_η. Note that P_η and $P_{\bar{\eta}}$ thus contain all possible (refined) transportation processes. ■

4.1.3 Allocation of System Function onto System Form

In hetero-functional graph theory, the allocation of system processes to system resources is captured in the "design equation," as shown below [8]:

$$P = J_S \odot R \tag{4.6}$$

where J_S is the system knowledge base, and $\odot$ is "matrix boolean multiplication" [8, 24].

Definition 4.10 – System Knowledge Base [8, 10, 14, 17, 20, 22][7]: A binary matrix J_S of size $\sigma(P) \times \sigma(R)$ whose element $J_S(w, v) \in \{0, 1\}$ is equal to one when action $e_{wv} \in \mathcal{E}$ (in the SysML sense) exists as a system process $p_w \in P$ being executed by a resource $r_v \in R$. The $\sigma()$ notation gives the size of a set. ■

In other words, the system knowledge base forms a bipartite graph [14] between the set of system processes and the set of system resources.

It is important to note that the system knowledge base views the set of system processes at the reference architecture level and the set of system resources at the instantiated architecture level [17]. As mentioned in Chapter 3, the mapping of the instantiated functional architecture to the instantiated physical architecture is *necessarily* one-to-one in large flexible engineering systems. In using the system processes defined at the reference architecture, the system knowledge base provides further insight into process redundancy [8, 17, 28].

$$\mathcal{R}_w = \sum_{v}^{\sigma(R)} J_S(w, v) \tag{4.7}$$

Furthermore, it is possible to aggregate resources R into aggregated resources $\overline{R}$ [8, 10, 17, 22, 28]:

[7]The system knowledge base was first introduced in [20] simply as a binary matrix and was later named as a knowledge base in [8, 10, 17, 22] in recognition of its use in the Axiomatic Design literature [46].

$$\overline{R} = \Xi \circledast R \tag{4.8}$$

where Ξ is an aggregation matrix and $\circledast$ is the aggregation operator.

Definition 4.11 – Aggregation Operator $\circledast$ [8, 10, 17, 22, 28]

$$C(i,k) = \bigcup_j a(i,j) \odot b(j,k) = A \circledast B \tag{4.9}$$

■

Some aggregated resources $\overline{R}$ can be considered *flexible* in that they are capable of executing more than one system process [8, 17, 28].

$$\mathcal{F}_v = \sum_w^{\sigma(P)} J_S(w,v) \tag{4.10}$$

Process redundancy $\mathcal{R}$ and resource flexibility $\mathcal{F}$ are valuable measures that directly impact a system's reconfigurability and resilience [14, 19].

The design equation is also applicable to the different types of processes and resources [8, 10, 14, 20, 22]:

$$P_\mu = J_M \odot M \tag{4.11}$$

$$P_\eta = J_H \odot R \tag{4.12}$$

$$P_\gamma = J_\gamma \odot R \tag{4.13}$$

where J_M is the transformation knowledge base, J_H is the transportation knowledge base, and J_γ is the holding knowledge base. The refined transportation knowledge base $J_{\bar{H}}$ is constructed as follows [8, 11, 14, 16, 18, 23][8]:

$$J_{\bar{H}} = \left[J_\gamma \otimes \mathbb{1}^{\sigma(P_\eta)}\right] \cdot \left[\mathbb{1}^{\sigma(P_\gamma)} \otimes J_H\right] \tag{4.14}$$

where $\otimes$ is the Kronecker product, and $\mathbb{1}^n$ is a ones-vector of length n.

Note that the mathematical structure of $J_{\bar{H}}$ effectively stratifies several copies of J_H so that system transportation processes are differentiated by the holding processes in P_γ. These holding processes are introduced in either of the three scenarios, as mentioned previously regarding the set of refined transportation processes.

[8]While more recent references represent $J_{\bar{H}}$ in matrix form, the original formula in [8] based upon scalars remains useful in large-scale computational applications where memory is limited.

In previous two-operand system literature, a more practical approach was used, such that the operand specific refined transportation knowledge bases were simply concatenated. The resources that appear in both systems were merged after concatenation. This approach is equally correct, potentially quicker for two-operand systems, but not practical for multi-operand systems. For example, the concatenation of transportation knowledge bases was used in electrified transportation systems [16, 18, 47] and microgrid-enabled production systems [40]. In greater detail, in [16, 47], the holding knowledge base differentiated between several different modes of travel within an electrified transportation infrastructure. The transportation system knowledge base for a system that transports *multiple* operands is a concatenation of the transportation knowledge bases for each holding process.

$$J_{\bar{H}} = \begin{bmatrix} J_{\bar{H}1} \\ \vdots \\ J_{\bar{H}n} \end{bmatrix} \tag{4.15}$$

In either case, the system knowledge base J_S is [8, 10, 14, 20, 22]:

$$J_S = \left[\begin{array}{c|c} J_M & \mathbf{0} \\ \hline \multicolumn{2}{c}{J_{\bar{H}}} \end{array} \right] \tag{4.16}$$

Hetero-functional graph theory also differentiates between the *existence* and the *availability* of physical capabilities in the system [8, 23]. While the former is described by the system knowledge base the latter is captured by the system constraints matrix.

Ontological Remark 4.4 *Building on the meta-architecture of the system processes and system resources, the mapping of system function to system form at the meta-level is indicated by the operations in Figure 4.6 or by the swim lanes in Figure 4.7.*

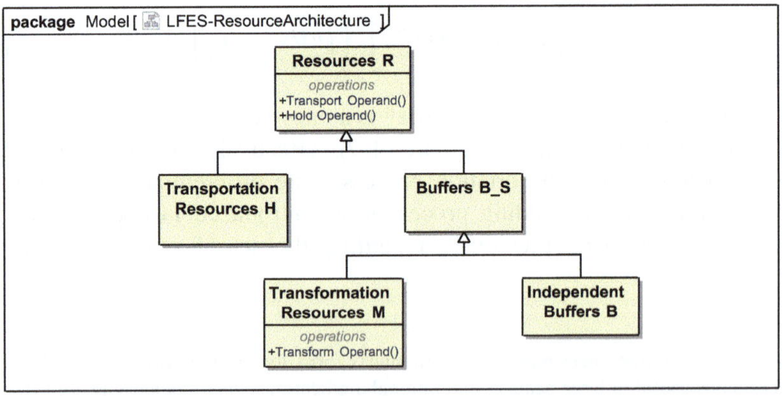

Fig. 4.6 A SysML Block Diagram: the meta-architecture of the allocated architecture of an LFES from a system form perspective

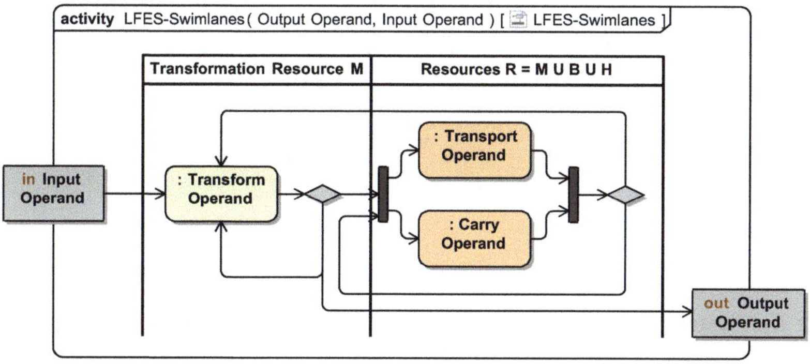

Fig. 4.7 A SysML Activity Diagram: the meta-architecture of the allocated architecture of an LFES from a system function perspective

Transformation resources are able to execute transformation processes while all resources are able to execute transportation processes. ■

Definition 4.12 – System Constraints Matrix [8, 10, 14, 17, 20, 22][9]: A binary matrix K_S of size $\sigma(P) \times \sigma(R)$ whose element $K_S(w, v) \in \{0, 1\}$ is equal to one when a constraint eliminates event e_{wv} from the event set. ■

The system constraints matrix is constructed analogously to the system knowledge base [8, 10, 14, 20, 22].

$$K_S = \left[\begin{array}{c|c} K_M & \mathbf{0} \\ \hline \multicolumn{2}{c}{K_{\bar{H}}} \end{array} \right] \tag{4.17}$$

In this regard, the system constraints matrix has a similar meaning to graph percolation [3, 34] and temporal networks [26].

A system's physical capabilities are quantified as structural degrees of freedom.

Definition 4.13 – Structural Degrees of Freedom [8, 10, 14, 17, 20, 22][10]: The set of independent actions $\mathcal{E}_S$ that completely defines the available processes in a large flexible engineering system. Their number is given by:

[9]The system constraints matrix was first introduced in [20] simply as a binary matrix and was later named in [8, 10, 14, 17, 22]. In some hetero-functional graph theory references, it is called the scleronomic constraints matrix or the sequence-independent constraints matrix to indicate that it is applied one capability at a time.

[10]The concept of structural degrees of freedom was first introduced in [20]. In some hetero-functional graph theory references, it is called scleronomic or sequence-independent degrees of freedom to indicate that they address capabilities one at a time.

$$DOF_S = \sigma(\mathcal{E}_S) = \sum_{w}^{\sigma(P)}\sum_{v}^{\sigma(R)} [J_S \ominus K_S](w,v) \tag{4.18}$$

$$= \sum_{w}^{\sigma(P)}\sum_{v}^{\sigma(R)} A_S(w,v) \tag{4.19}$$

$$= \langle J_S, \bar{K}_S \rangle_F \tag{4.20}$$

where the system constraints matrix limits the availability of degrees of freedom in the system knowledge base to create the system concept A_S. Note that the transformation degrees of freedom DOF_M and the refined transportation degrees of freedom DOF_H are calculated similarly [8, 10, 20, 22]:

$$DOF_M = \sum_{j}^{\sigma(P_\mu)}\sum_{m}^{\sigma(M)} [J_M \ominus K_M](j,m) \tag{4.21}$$

$$DOF_H = \sum_{u}^{\sigma(P_{\bar{\eta}})}\sum_{v}^{\sigma(R)} \left[J_{\bar{H}} \ominus K_{\bar{H}}\right](u,v) \tag{4.22}$$

■

Example 4.3 The term structural degrees of freedom is best viewed as a generalization of kinematic degrees of freedom (or generalized coordinates) [8, 10, 12, 14, 20, 22]. Their number is given by [42]:

$$DOF = n_d * n_l - n_k \tag{4.23}$$

where n_l is the number of links, $n_d = 6$ is the number of primitive coordinates, and n_k is the number of applied scleronomic (i.e., time independent) constraints that confine motion. For example, consider the two-bar linkage shown in Figure 4.8.

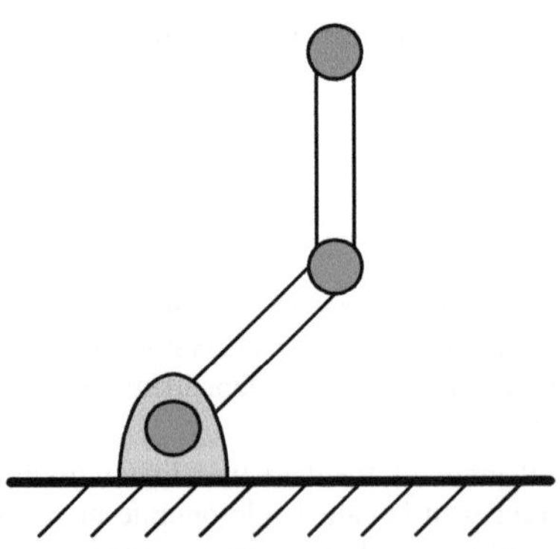

Fig. 4.8 A two-bar linkage system

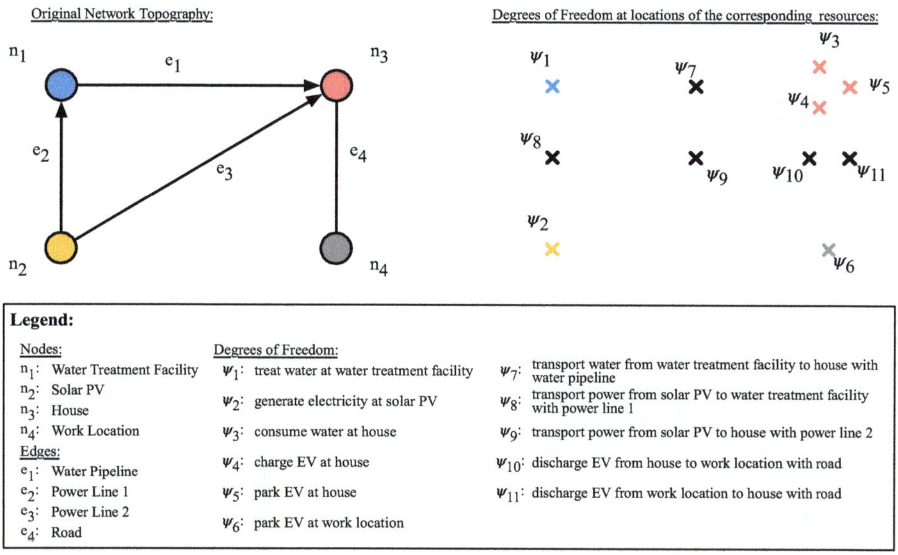

Fig. 4.9 Degrees of freedom in the example network: a visual comparison of the original network topography on the left, and the system degrees of freedom on the right

$n_l = 2$ and $n_k = 2 * 5 = 10$. Consequently $DOF = 2$ as expected. If any individual constraint is assumed to affect only a single combination of link and coordinate, then Equation 4.23 can be written as [8, 10, 14, 17, 20, 22]:

$$DOF = \sum_{i}^{n_d} \sum_{j}^{n_l} [J_S \ominus K_S](i, j) = \sum_{i}^{n_d} \sum_{j}^{n_l} A_S(i, j) \tag{4.24}$$

Consequently, in hetero-functional graph theory, mechanical links as elements of form are analogous to system resources and primitive coordinates as descriptors of function are analogous to system processes [8, 10, 12, 14, 20, 22]. ∎

Example 4.4 Building off Example 4.2 and returning to smart city applications, Figures 4.2 and 4.3 can be quantified for their system knowledge base J_S and system constraints matrix K_S. Figure 4.9 provides a visual overview of the system capabilities, presented at the location of their physical resources. Figure 4.10 provides the matrix representation of each of the knowledge bases.

The *transformation knowledge base* J_M maps the transformation processes to the transformation resources. The transformation knowledge base consequently has size $\sigma(P_\mu) \times \sigma(M) = 3 \times 3$, with three filled elements representing the transformation capabilities.

The *transportation knowledge base* J_H maps the transportation processes to the system resources, resulting in a matrix of size $\sigma(P_\eta) \times \sigma(R) = 16 \times 8$. Matrix J_H contains seven filled elements, representing the transportation capabilities.

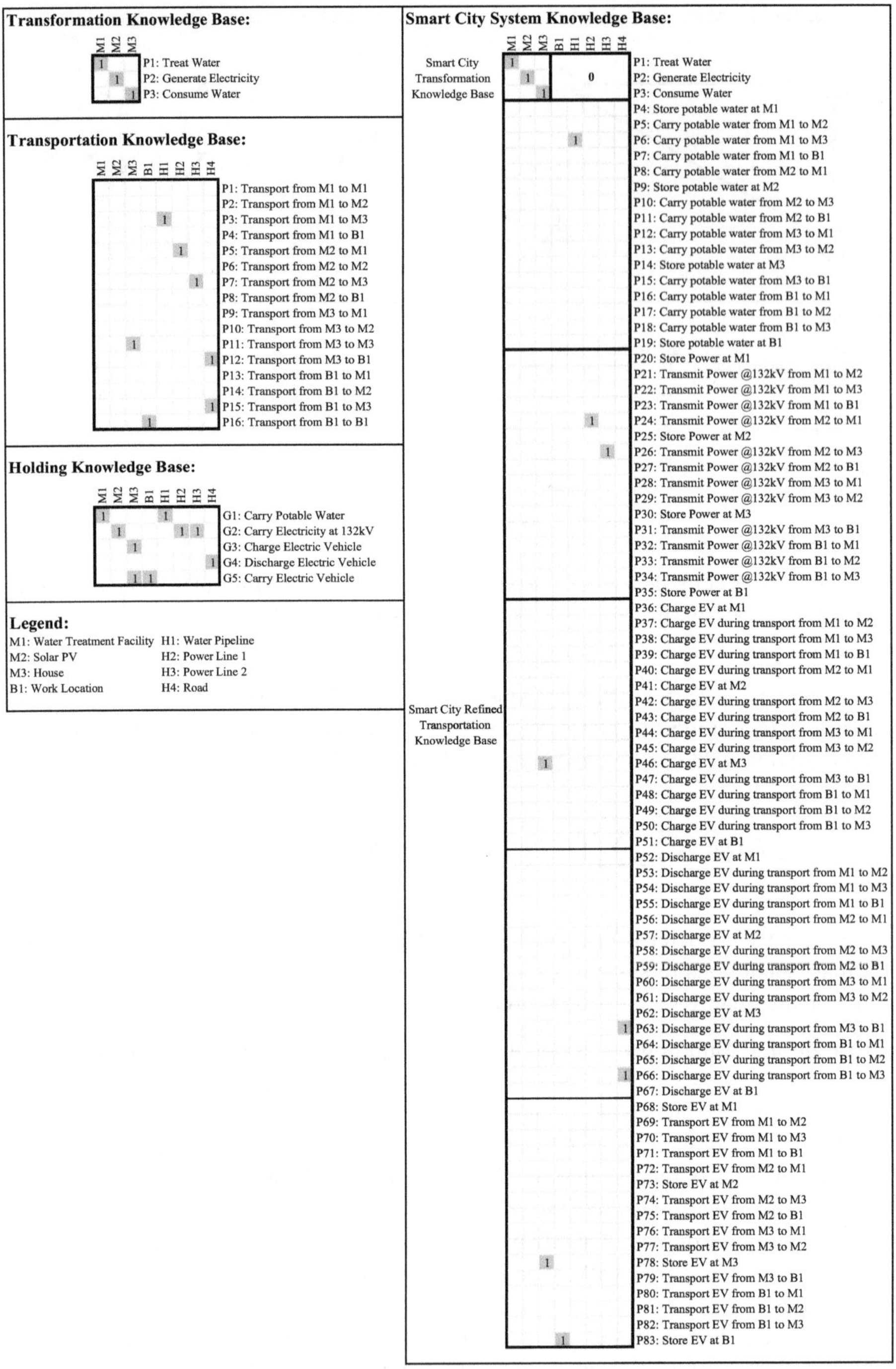

Fig. 4.10 Transformation, transportation, holding, and system knowledge bases corresponding to Figures 4.2 and 4.3

The *holding knowledge base* J_γ serves to refine the transportation processes. It distinguishes the operands, the way of holding operands, and the way of holding operands that could potentially change the state of the operand. The size of J_γ is $\sigma(P_\gamma) \times \sigma(R) = 5 \times 8$, with nine filled elements.

The *refined transportation knowledge base* $J_{\bar{H}}$ is calculated using Equation 4.14. Its size is $\sigma(P_{\bar{\eta}}) \times \sigma(R)$, and it contains eight filled elements.

The *system knowledge base* J_S then follows as a concatenation of J_M and $J_{\bar{H}}$, following Equation 4.16. The system knowledge base consequently has the size $\sigma(P) \times \sigma(R) = 83 \times 8$, with 11 filled elements that represent the system capabilities.

In this example, all capabilities are available and K_S is thus an all-zeros matrix with the same size as J_S. The number of available capabilities (i.e., degrees of freedom) is calculated to be 11 using Equation 4.18. In this case, the number of capabilities is equal to the number of actions in the activity diagram in Figure 4.3. ■

4.2 Hetero-functional Adjacency Matrix

In the previous subsection, system capabilities are described as sentences. In this subsection, the capabilities are connected into serial and parallel arrangements. This is similar to when sentences are connected together to form paragraphs and stories. The hetero-functional adjacency matrix is used to represent the logical order of physical capabilities in a system [14].

In SysML, both the block definition diagram and the activity diagram with swim lanes have the ability to represent the sequence of capabilities [25, 50]. Sequence in the block definition diagram is represented at the meta-architecture level, as the physical continuity applies to all engineering systems. Figure 4.11 represents the physical sequence constraints with associations in green. Sequence in the activity diagram uses the functional interactions between the actions to link the capabilities together. For example, Figure 4.3 means: "The solar PV panel generates electricity from solar irradiation. Power line 2 transmits electricity to the house."

In hetero-functional graph theory, the hetero-functional adjacency matrix A_ρ is introduced to represent the sequence of physical capabilities [14–16, 40, 48]. Much like the system concept A_S, the hetero-functional adjacency matrix A_ρ arises from a Boolean difference [14–16, 40, 48].

$$A_\rho = J_\rho \ominus K_\rho \tag{4.25}$$

where J_ρ is the system sequence knowledge base and K_ρ is the system sequence constraints matrix.

Definition 4.14 – System Sequence Knowledge Base [14–16, 40, 48][11]: A square binary matrix J_ρ of size $\sigma(R)\sigma(P) \times \sigma(R)\sigma(P)$ whose element $J_\rho(\chi_1, \chi_2) \in \{0, 1\}$

[11]One important change to hetero-functional graph theory over the years has been in the system sequence knowledge base. Originally, it was called the rheonomic knowledge base and defined as a

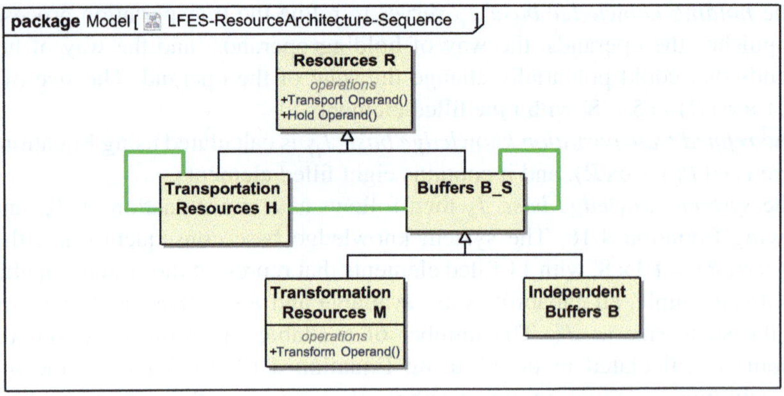

Fig. 4.11 A SysML Block Diagram: system sequence associations are added to the meta-architecture of the allocated architecture of an LFES from a system form perspective

is equal to one when string z_{χ_1,χ_2} exists, where index $\chi_i \in [1,\ldots,\sigma(R)\sigma(P)]$. It is calculated as:

$$J_\rho = A_S^V A_S^{VT} \tag{4.26}$$

$$J_\rho = \left[J_S \cdot \bar{K}_S\right]^V \left[J_S \cdot \bar{K}_S\right]^{VT} \tag{4.27}$$

where $()^V$ is shorthand for vectorization (i.e. vec()). ■

Definition 4.15 – System Sequence Constraints Matrix [14–16, 40, 48][12]: A square binary constraints matrix K_ρ of size $\sigma(R)\sigma(P)\times\sigma(R)\sigma(P)$ whose elements $K(\chi_1,\chi_2) \in \{0,1\}$ are equal to one when string $z_{\chi_1\chi_2} = e_{w_1v_1}e_{w_2v_2} \in \mathcal{Z}$ is eliminated. ■

Note that J_ρ creates all potential sequences of the capabilities in A_S. However, there are only certain pairs of capabilities that are feasible. The system sequence constraints matrix K_ρ serves to eliminate the infeasible pairs.

Sequence dependent constraints in K_ρ may be imposed on either functional interactions and/or physical interfaces. Consequently, several considerations must be taken into account in calculating K_ρ. In the case of functional interactions within an instantiated functional architecture, the *absence* of an arrow between any two

binary matrix of size $\sigma^2(P)\times\sigma^2(R)$ to mirror its sequence-independent counterpart [8, 10, 20, 22]. Later works have adopted the definition presented above as a methodological development.

[12]Similarly, the system sequence constraints matrix was first called the rheonomic knowledge base and defined as a binary matrix of size $\sigma^2(P) \times \sigma^2(R)$. Later works have adopted the definition presented above as a methodological development.

Table 4.3 Types of sequence-dependent production degree of freedom measures [11, 14, 48]

Type	Measure	Process	Resource	Sequence-dependent knowledge base	Constraint matrix	Perpetual constraint	Measure function
I	$DOF_{MM\rho}$	$P_\mu P_\mu$	M, M	$J_{MM\rho} = [J_M \cdot \bar{K}_M]^V [J_M \cdot \bar{K}_M]^{VT}$	$K_{MM\rho}$	$m_1 = m_2$	$\langle J_{MM\rho}, \bar{K}_{MM\rho}\rangle_F$
II	$DOF_{MH\rho}$	$P_\mu P_\eta$	M, R	$J_{MH\rho} = [J_M \cdot \bar{K}_M]^V [J_H \cdot \bar{K}_H]^{VT}$	$K_{MH\rho}$	$m_1 - 1 = (u_1 - 1)/\sigma(B_S)$	$\langle J_{MH\rho}, \bar{K}_{MH\rho}\rangle_F$
III	$DOF_{HM\rho}$	$P_\eta P_\mu$	R, M	$J_{HM\rho} = [J_H \cdot \bar{K}_H]^V [J_M \cdot \bar{K}_M]^{VT}$	$K_{HM\rho}$	$m_1 - 1 = (u_1 - 1)\&\sigma(B_S)$	$\langle J_{HM\rho}, \bar{K}_{HM\rho}\rangle_F$
IV	$DOF_{HH\rho}$	$P_\eta P_\eta$	R, R	$J_{HH\rho} = [J_H \cdot \bar{K}_H]^V [J_H \cdot \bar{K}_H]^{VT}$	$K_{HH\rho}$	$(u_1 - 1)\%\sigma(B_S) = (u_2 - 1)/\sigma(B_S)$	$\langle J_{HH\rho}, \bar{K}_{HH\rho}\rangle_F$
ALL	DOF_ρ	PP	R, R	$J_\rho = [J_S \cdot \bar{K}_S]^V [J_S \cdot \bar{K}_S]^{VT}$	K_ρ	All of the above	$\langle J_{S\rho}, \bar{K}_{S\rho}\rangle_F$

allocated actions in a SysML activity diagram imposes a set of functional constraints within K_ρ. Similarly, in the case of physical interfaces within an instantiated physical architecture, the *absence* of an association link in a SysML block diagram similarly imposes a set of physical constraints within K_ρ (without regard for system function). Similarly, *domain-specific* sequence dependent constraints in K_ρ may be imposed directly at the reference architecture level; be they from arrows in the SysML activity diagram or from association links in the SysML block diagram. Such design pattern constraints effectively impose topological exchanges of matter, energy, information, people, or money [6].[13] Finally, in the absence of either an explicitly stated physical or functional architecture be it at the instantiated system or reference level, the system sequence constraints matrix K_ρ must still address continuity as a meta-level property of physical architectures (in general). In other words, given two capabilities, the output of the first must occur at the location of the input of the second. These constraints are described quantitatively in Table 4.3 [14]. In such a case, the construction of the system sequence constraints matrix requires the tracking of all four constraints for each of the pairs of capabilities in the system. A straightforward way of calculating this matrix is a scalar implementation using FOR loops while adhering to the following relationships of indices $\chi = \sigma(P)(v-1)+w$. For a transformation process p_w, $w = j\ \forall j = [1 \ldots \sigma(P_\mu)]$. For a transportation process p_w, $w = [\sigma(P_\eta)(g-1)+u]+\sigma(P_\mu)\ \forall g = [1 \ldots \sigma(P_\gamma)], \forall u = [1 \ldots \sigma(P_\eta)]$ [11, 14, 48].

A system's sequences are quantified as sequence-dependent degrees of freedom.

[13] In [4], Crawley distinguishes between topological connections and spatial relationships as two types of interactions. Because spatial relationships are effectively a type of requirement and are not typically addressed in the network sciences, hetero-functional graph theory neglects their treatment here as well.

Definition 4.16 – Sequence-Dependent Degrees of Freedom [14–16, 40, 48][14]**:** The set of independent pairs of actions $z_{\chi_1\chi_2} = e_{w_1v_1}e_{w_2v_2} \in \mathcal{Z}$ of length 2 that completely describe the system language. The number is given by:

$$DOF_\rho = \sigma(\mathcal{Z}) = \sum_{\chi_1}^{\sigma(R)\sigma(P)} \sum_{\chi_2}^{\sigma(R)\sigma(P)} [J_\rho \ominus K_\rho](\chi_1, \chi_2) \tag{4.28}$$

$$= \sum_{\chi_1}^{\sigma(R)\sigma(P)} \sum_{\chi_2}^{\sigma(R)\sigma(P)} [A_\rho](\chi_1, \chi_2) \tag{4.29}$$

■

For systems of substantial size, the size of the hetero-functional adjacency matrix may be challenging to process computationally. However, the matrix is generally very sparse. Therefore, projection operators are used to eliminate the sparsity by projecting the matrix onto a one's vector [16, 40]. This is demonstrated below for J_S^V and A_ρ:

$$\mathbb{P}_S J_S^V = \mathbb{1}^{\sigma(\mathcal{E}_S)} \tag{4.30}$$

$$\mathbb{P}_S A_\rho \mathbb{P}_S^T = \widetilde{A}_\rho \tag{4.31}$$

where $\mathbb{P}_S$ is a (non-unique) projection matrix for the vectorized system knowledge base and the hetero-functional adjacency matrix [16, 40].

Note that the number of sequence dependent degrees of freedom for the projected hetero-functional adjacency matrix can be calculated as:

$$DOF_\rho = \sigma(\mathcal{Z}) = \sum_{\psi_1}^{\sigma(\mathcal{E}_S)} \sum_{\psi_2}^{\sigma(\mathcal{E}_S)} [\widetilde{A}_\rho](\psi_1, \psi_2) \tag{4.32}$$

where $\psi \in [1, \ldots, \sigma(\mathcal{E}_S)]$.

Example 4.5 This example continues where Examples 4.2 and 4.4 left off. Recall that Example 4.2 derived the sets of the 4-node example network, and that Example 4.4 continued by calculating the knowledge bases. This example completes the structural model for the 4-node network by calculating the hetero-functional adjacency matrix A_ρ. First, the system sequence knowledge base J_ρ is calculated using Equation 4.26. Recall that no constraints were imposed on the knowledge base

[14] Similarly, in earlier works, the sequence-dependent degrees of freedom measure was first called the rheonomic degrees of freedom and defined over a binary matrix of size $\sigma^2(P) \times \sigma^2(R)$. Despite this methodological difference, it is easy to show that the number of sequence dependent degrees of freedom is the same regardless of the choice of calculation method.

	1	2	3	4	5	6	7	8	9	10	11
1							1				
2								1	1		
3											
4				1	1					1	
5				1	1					1	
6						1					1
7			1								
8	1										
9			1	1							
10						1					1
11				1	1					1	

Legend:

1: treat water at water treatment facility
2: generate electricity at solar PV
3: consume water at house
4: charge EV at house
5: park EV at house
6: park EV at work location
7: transport water from water treatment facility to house with water pipeline
8: transport power from solar PV to water treatment facility with power line 1
9: transport power from solar PV to house with power line 2
10: discharge EV from house to work location with road
11: discharge EV from work location to house with road

Fig. 4.12 Projected hetero-functional adjacency matrix $\tilde{A}_\rho$ for Example 4.5 (row and column sparsity have been eliminated)

of the 4-node network, and consequently, size of matrix A_S is $\sigma(P)\times\sigma(R) = 83\times8$, with 11 filled elements. The resulting size of matrix J_ρ is $\sigma(R)\sigma(P)\times\sigma(R)\sigma(P) = 664 \times 664$, with $11^2 = 121$ filled elements.

Second, each of the elements in the system sequence knowledge base can be checked for its adherence to the physical sequence constraints as defined in Table 4.3. Additionally, the sequences must adhere to the functional sequences as defined in the activity diagram in Figure 4.3 on Page 42. By directly checking the sequences for adherence to the constraints, the calculation of the full constraints matrix K_ρ is avoided, and the hetero-functional adjacency matrix A_ρ is calculated directly. Matrix A_ρ has size $\sigma(R)\sigma(P) \times \sigma(R)\sigma(P) = 664 \times 664$, with 20 filled elements.

Last, the matrix A_ρ is projected and visualized. The projection operation eliminates the row and column sparsity, so as to reduce the size of the hetero-functional adjacency matrix (using Equation 4.31). This is especially important for more complex systems. The projected hetero-functional adjacency matrix $\widetilde{A}_\rho$ now has size $\sigma(\mathcal{E}_S) \times \sigma(\mathcal{E}_S) = 11 \times 11$, still with 20 filled elements. The matrix is presented in Figure 4.12, and presented as a network in Figure 4.13. ■

Ontological Remark 4.5 *It is important to recognize that the hetero-functional adjacency matrix is entirely distinct from the (traditional) adjacency matrices often used in graph theory. While the latter often represents the nodes as some element of form (i.e., a resource) and edges as the flows of matter, energy, or information between them, the hetero-functional adjacency matrix has structural degrees of freedom (or capabilities) as nodes and edges to represent feasible logical sequences. This is an essential difference. While a traditional adjacency matrix assigns only one node to a resource regardless of its functionality, a hetero-functional adjacency matrix assigns a node to each capability that each resource can perform. A traditional adjacency matrix uses edges to represent transportation resources and implicitly assumes a one-to-one relationship with the transportation processes that*

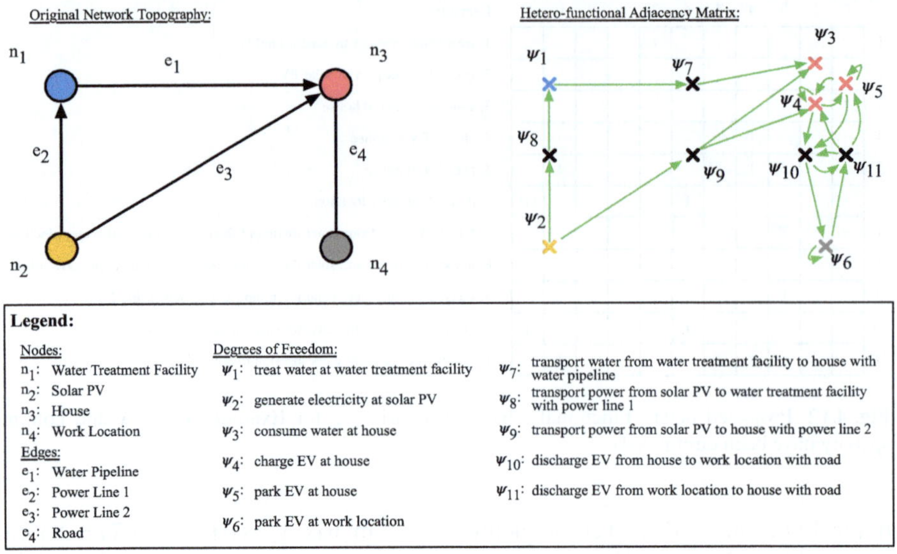

Fig. 4.13 Degrees of freedom in the example network: a visual comparison of the original network topography on the left, and the hetero-functional adjacency matrix on the right

they execute. A hetero-functional adjacency matrix, in contrast, assigns nodes to transportation capabilities. This directly facilitates an understanding of the redundancy of each transportation process. The explicit treatment of both the functional and physical architectures and their embedded process redundancy and resource flexibility become an integral part of understanding the structure and behavior of integrated smart city infrastructures. ■

4.3 Controller Agency Matrix

The controller agency matrix serves to differentiate between two systems of equivalent capabilities but different control structure [8–10, 22]. Consider Figure 4.14 [8]. Both systems have equivalent capabilities; they are able to do the same processes on the same resources. The system on the right, however, requires a centralized controller to perform these capabilities; while the system on the left can do so without the need for such a controller. Ontologically speaking, the controller agency matrix is introduced to ensure the lucidity of the model. In the case of centralized controllers, the set of system resources R is assumed to now include cyber-resources Q. $R = M \cup B \cup H \cup Q$ [8–10, 22]. Furthermore, the previously identified resources M, B, and H constitute the system's physical resources $R_P = M \cup B \cup H$.

Definition 4.17 – Cyber-Resource [8–10, 22]: A resource $r \in R$ is a cyber-resource (e.g., controller or decision-making agent) $q \in Q = Q_D \cup Q_I$ iff it

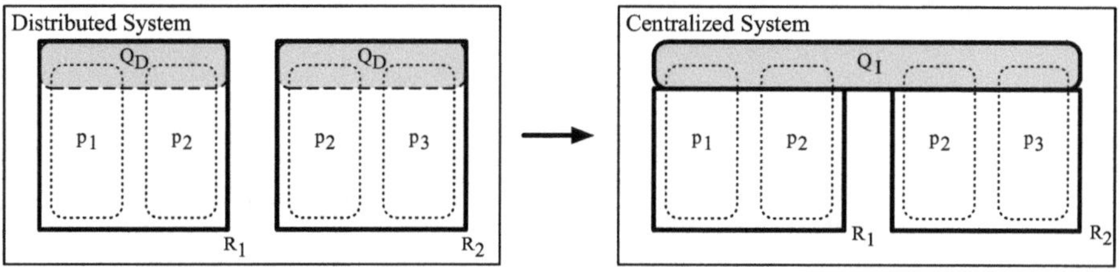

Fig. 4.14 Capabilities with cyber-resources. The distributed system on the left has embedded (dependent) controller Q_D, and the centralized system on the right has an independent controller Q_I [8]

is capable of controlling how physical resources perform their system processes. *Dependent* cyber-resources Q_D are integral parts of physical resources (as in the case of embedded controllers). Independent cyber-resources Q_I are stand-alone and external to the physical resources (e.g., centralized controllers and decision-making entities). For the sake of simplicity, each cyber-resource is assumed to be able to execute a single control or decision algorithm $p_Q \in P_Q$. Cyber-resources that can execute multiple decisions are called aggregated cyber-resources $\overline{Q} = \Xi \circledast Q$.[15] ■

The control structure of an engineering system describes the scope of control or jurisdiction that a cyber-resource has over a physical resource. If a physical resource has one or more dependent cyber-resources, then it has its own control or decision-making agency and is usually able to behave autonomously or in coordination with other resources [15, 23]. Other physical resources must instead rely on independent cyber-resources in order to effectively control their capabilities.

Ontological Remark 4.6 *The above paragraphs imply a revision to the meta-architecture of a large flexible engineering system. The form primitives of hetero-functional graph theory now include a set of cyber-resources Q. Consequently, the classification depicted in Figure 4.4 is revised to the one depicted in Figure 4.15. Cyber-resources (i.e., controllers and decision-making agents) are added as a class. Dependent cyber-resources are linked to their associated physical resources with decomposition links. However, for modeling simplicity, this is rarely shown unless there is a need to decompose the resource into its physical and cyber parts. In such a case* $Q_D = \emptyset, Q = Q_I$. *Instead, physical resources with their own control or decision-making agency may be viewed as having an association link to itself. Independent cyber-resources are linked to the physical resources that they control with association links. Meanwhile, the primitives of the meta-architecture of system function is a set of processes that now include control and decision algorithms* P_Q. *Consequently, Figure 4.5 is revised to the one shown in Figure 4.16.* ■

[15]The definition presented above is a generalization of the one found in earlier works [8–10, 22] where dependent cyber-resources were neglected and independent cyber-resources were called centralized controllers.

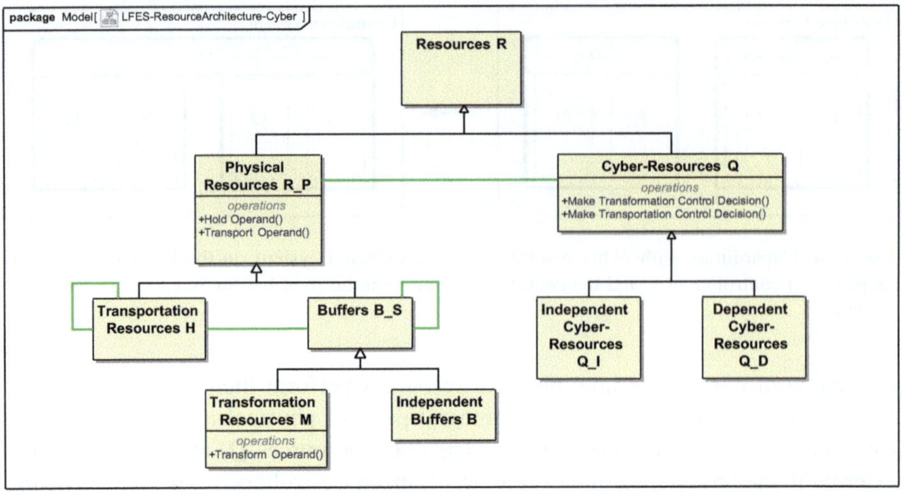

Fig. 4.15 A SysML Block Diagram: the meta-architecture of the system form of an LFES with cyber-resources

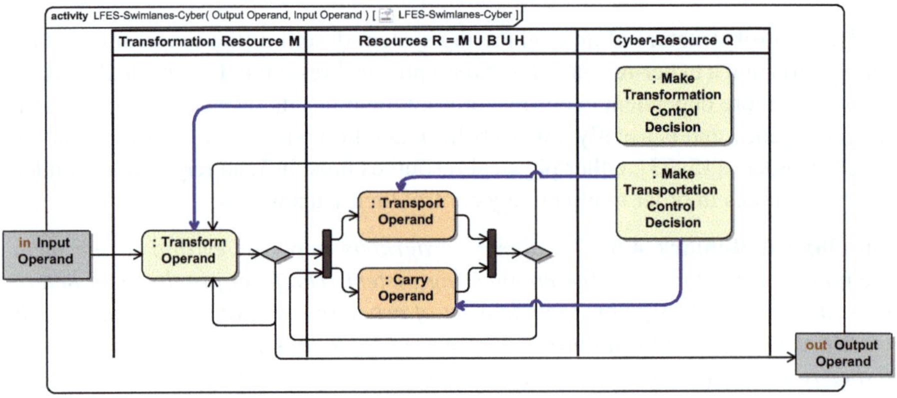

Fig. 4.16 A SysML Activity Diagram: the meta-architecture of the system function of an LFES with control and decision-making algorithms

In hetero-functional graph theory, the controller agency matrix A_Q defines the control structure of a system.

Definition 4.18 – Controller Agency Matrix [8–10, 22]: A binary matrix A_Q of size $\sigma(R_P) \times \sigma(R)$, whose element $A_Q(v_1, v_2)$ is equal to one when the resource $r_{v_2} \in R$ has control jurisdiction over the physical resource $r_{v_1} \in R_P$. When $A_Q(v_1, v_1) = 1$, a physical resource $r_{v_1} \in R_P$ is said to control itself (e.g., by an embedded controller). When $A_Q(v_1, v_2) = 1$ and $v_1 \neq v_2$ then the independent cyber-resource $r_{v_2} \in Q$ is required (i.e., as a centralized controller) to control the physical resource r_{v_1}. In the absence of independent cyber-resources,

$Q = \emptyset$, $A_Q = I^{\sigma(R_P)}$. In an engineering system with a *perfectly* centralized control system (i.e., one with a single centralized controller controlling multiple physical resources), $A_Q = [I^{\sigma(R_P)}, \mathbb{1}^{\sigma(R_P)}]$. In an engineering system with only *physical agents* (i.e., one agent for every physical resource), $A_Q = [I^{\sigma(R_P)}, I^{\sigma(R_P)}]$. In such a case, the physical agents are often treated as *dependent cyber-resources*, and the cyber-physical resources $\overline{R}_P = A_Q \circledast R$ are used in place of the physical resources R_P. In all other cases, $A_Q = [I^{\sigma(R_P)}, \overline{A}_Q]$ where $\overline{A}_Q$ is the independent controller agency matrix. $\overline{A}_Q$ is a bipartite graph that shows the jurisdiction of independent cyber-resources Q over physical resources R_P. ■

Ontological Remark 4.7 *Hetero-functional graph theory introduces the controller agency matrix in order to introduce ontological lucidity; recognizing that cyber and physical interactions are fundamentally different [1, 7, 35]. Flows of power are fundamentally two way; power cannot be transferred from one entity to another without affecting the state of both entities. Flows of matter are either one way or two way, depending on whether the matter-flow is modeled to have an associated power flow. Meanwhile, in most engineering applications information is assumed to be one-way, aphysical, and of negligible power transfer. When information is passed from one entity to another, only the recipient but not the sender is affected. In quantum information processing, however, the information regains its physical characteristics. The recipient and sender are affected by the transfer of information [33]. The controller agency matrix shows* informatic *interactions between the resources R and the physical resources R_P. This is contrasted to the hetero-functional adjacency matrix which shows* physical *interactions between physical capabilities (i.e., structural degrees of freedom).*

When the engineering system has independent cyber-resources, the controller agency matrix is necessary to describe the allocation of system processes to system resources. The "design equation" in Equation 4.6 is then generalized to become:

$$P = J_S \odot (A_Q \circledast R) \tag{4.33}$$

where $R = R_P \cup Q$ and $\circledast$ is the aggregation operator.

In this regard, hetero-functional graph theory recognizes that physical and cyber resources are fundamentally different in nature. While the former has one or more associated capabilities, the latter is only able to affect how these system capabilities perform. The interactions described in the controller agency matrix are fundamentally different from those found in the system concept A_S and the hetero-functional adjacency matrix A_ρ. Mathematically speaking, A_Q specifies a relationship between two resources (rather than between processes and resources in the case of A_S, or between capabilities in the case of A_ρ). Conceptually, A_Q specifies control jurisdiction whereas A_S represents allocation of function to form and A_ρ describes the flow of matter and energy between capabilities. These distinctions make hetero-functional graph theory inherently cyber-physical. ■

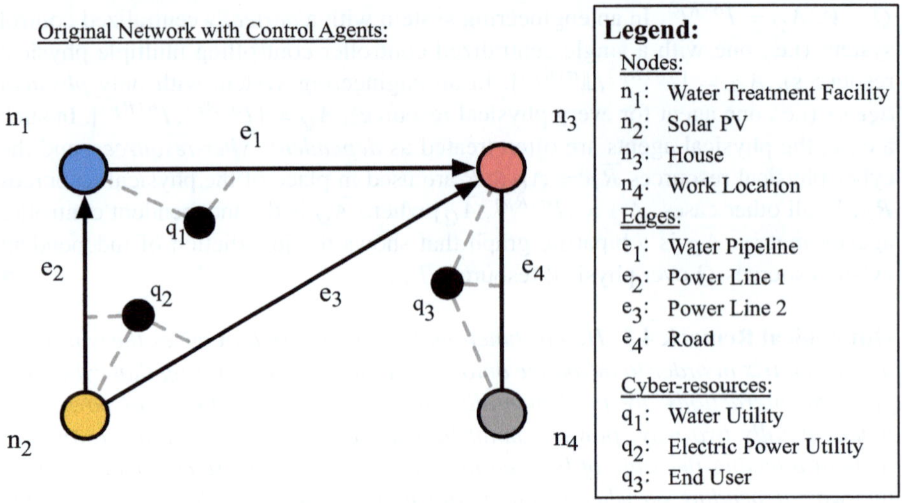

Fig. 4.17 Cyber-resources in the Example Network: independent cyber resources have jurisdiction over physical resources. Cyber-physical interfaces are indicated with grey dashed edges

Example 4.6 Building off Example 4.2, the network in Figure 4.1 can be enhanced to include independent cyber-resources as shown in Figure 4.17. The SysML block diagram in Figure 4.2 can be enhanced similarly, as shown in Figure 4.18. The electric power utility, the water utility, and the end users are all considered cyber-resources for their ability to take control actions or make decisions. Associations are then introduced between cyber-resources and their associated (physical) resources to visualize the control structure. SysML uses associations to represent both physical and informatic interfaces whereas hetero-functional graph theory differentiates between the two. All resources without an independent cyber-resource are assumed to have a dependent cyber-resource which has not been shown for modeling simplicity. $Q_D = \emptyset, Q = Q_I$. Similarly, the self-associations reflecting a physical resource's ability to control itself have not been shown to maintain the graphical aesthetic. These associations are used to construct the controller agency matrix in Figure 4.19. The physical resource's self-associations appear as an identity matrix block. The associations between the physical and independent cyber-resources appear as filled elements in the remaining columns. ■

4.4 Controller Adjacency Matrix

The controller adjacency matrix serves to express the interactions between cyber-resources. Smart city infrastructure systems are controlled by a diverse set of actors, each of whom have jurisdiction over parts of the smart city (e.g., an operand-layer

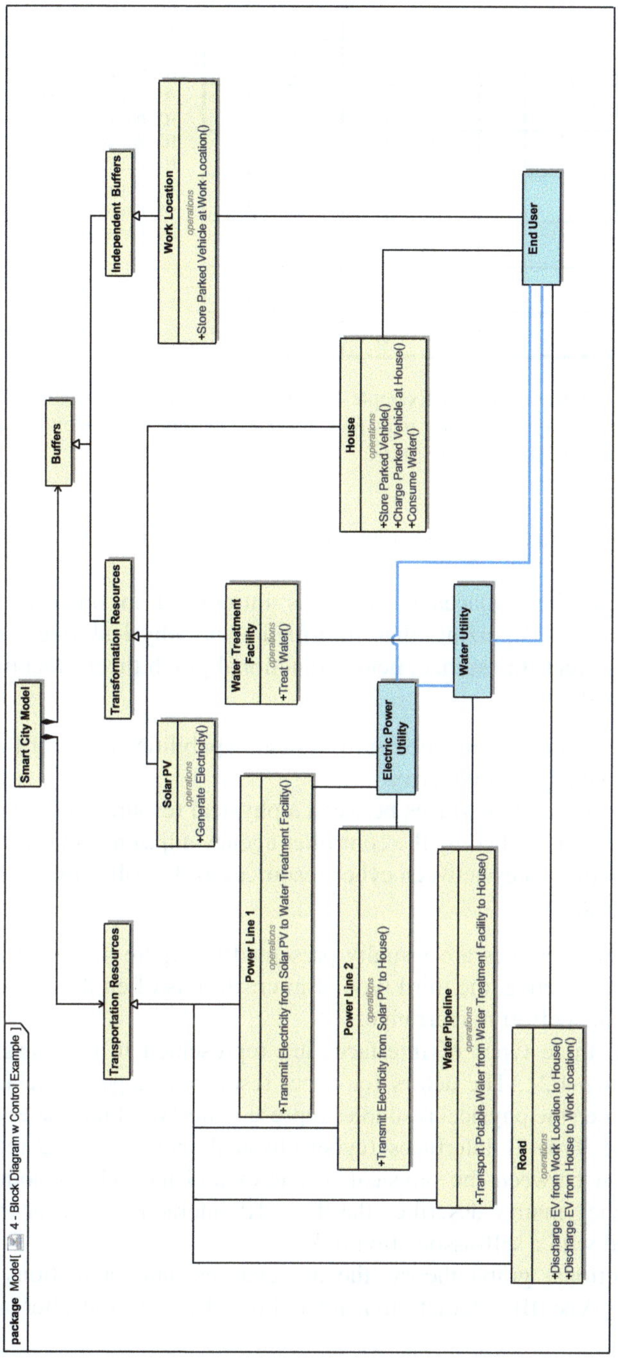

Fig. 4.18 SysML Block Definition Diagram for Example 4.6. This block diagram extends the block diagram from Figure 4.2 to include the Control Agents, who have control authority over resources via (colored) associations

	M1	M2	M3	B1	H1	H2	H3	H4	Q1	Q2	Q3
M1	1								1		
M2		1								1	
M3			1								1
B1				1							1
H1					1				1		
H2						1				1	
H3							1			1	
H4								1			1

Legend:

M1: Water Treatment Facility
M2: Solar PV
M3: House
B1: Work Location
H1: Water Pipeline
H2: Power Line 1
H3: Power Line 2
H4: Road
Q1: Water Utility
Q2: Electric Power Utility
Q3: End User

Fig. 4.19 Controller agency matrix for Example 4.6. The block form matrix contains two blocks: (1) The left side: the identity matrix of size $\sigma(R_P) \times \sigma(R_P)$. (2) The right side: the independent controller agency matrix of size $\sigma(R_P) \times \sigma(Q)$

in a multi-operand infrastructure system). Operating the smart city requires these actors to interact.

Again, the controller adjacency matrix is introduced in order to maintain ontological lucidity; recognizing that informatic and physical interactions are fundamentally different. In essence, hetero-functional graph theory identifies three types of interfaces [8, 9]:

1. Type I: physical interactions between physical capabilities as described by the hetero-functional adjacency matrix.
2. Type II: cyber-physical interfaces between a physical resource and all resources cyber-resource as described by the controller agency adjacency matrix.
3. Type III: cyber-interfaces between cyber-resources as described by the controller adjacency matrix.

The three types of interfaces are visually presented in Figure 4.20 [8, 9]. From a reconfiguration perspective, the third type of interaction is often the most complex and requires the most effort to rearrange.

In SysML, all three types of interfaces are represented in the class diagram as shown in Figure 4.21. The actors (or cyber-resources) appear as classes. The associations between the physical resources represent the Type I interfaces (physical interactions). The Type II interfaces (cyber-physical interfaces) are represented by the association between the physical resources and the cyber-resources. The controller adjacency matrix describes the Type III interactions (cyber-interfaces) and is represented with a self-association link.

In hetero-functional graph theory, the associations and interactions between cyber-resources (Type III interactions) are defined by the controller adjacency matrix.

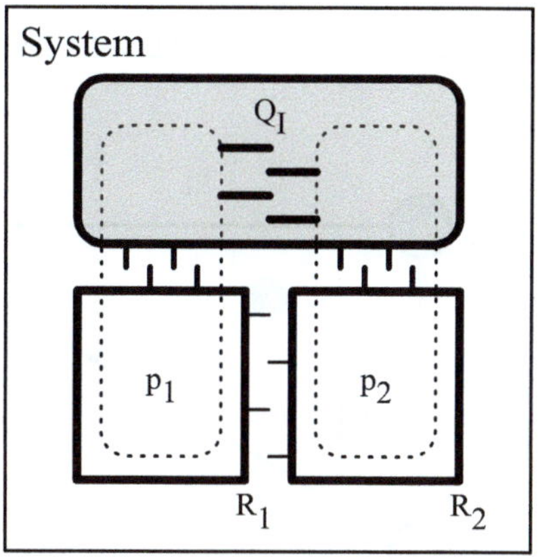

Fig. 4.20 Three types of interfaces between physical and cyber resources. Type I is between two physical resources. Type II is between a physical and a cyber-resource. Type III is between two cyber-resources (line thickness represents the complexity of interaction and separation) [8, 9]

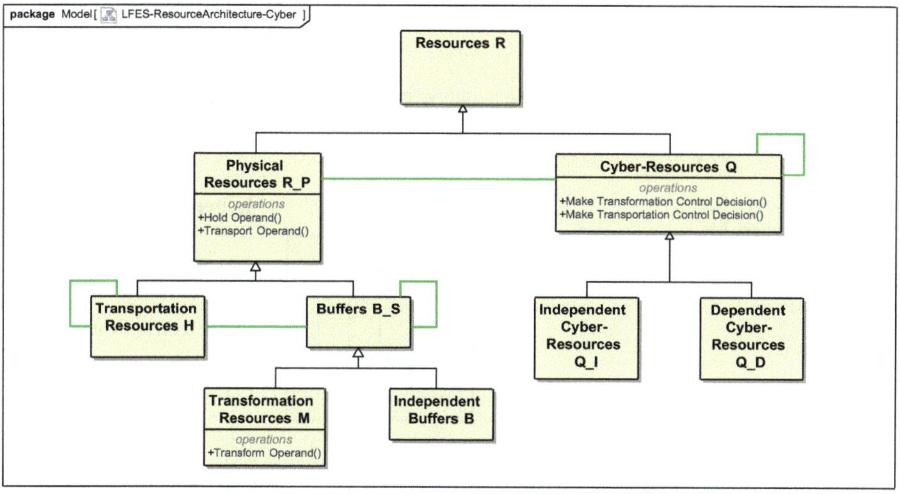

Fig. 4.21 A SysML Block Diagram: the meta-architecture of the system form of an LFES with cyber-resources and their adjacency

Definition 4.19 – Controller Adjacency Matrix: The controller adjacency matrix A_C is a binary matrix of size $\sigma(Q) \times \sigma(Q)$, whose element $A_C(v_1, v_2)$ is equal to one when the cyber-resource $q_{v_1} \in Q$ pass information to cyber-resource $q_{v_2} \in Q$. ■

The number of connections in the controller adjacency matrix provides insight into the complexity of the network and thus its operational challenges. The cross-layer interactions of physical resources (such as electric vehicle charging stations) often require cyber-resources to coordinate their operational activities as their actions may cause positive outcomes in one layer while causing undesired consequences in another.

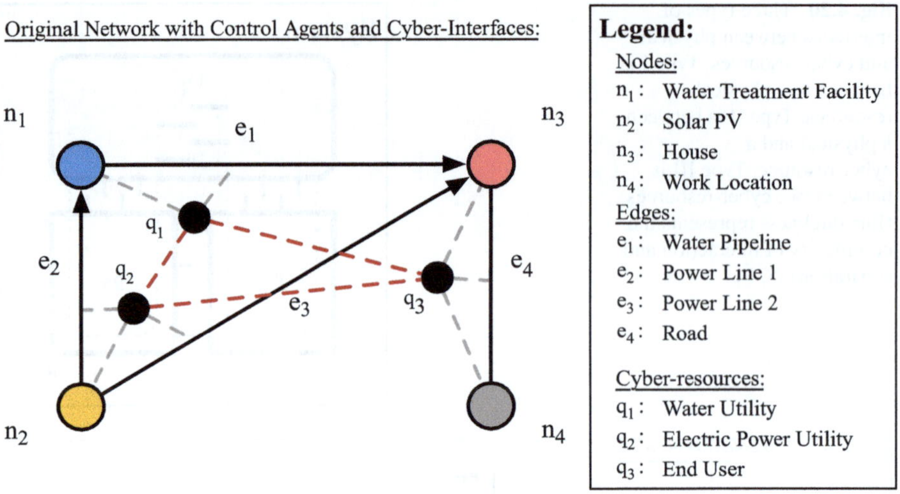

Fig. 4.22 Cyber-resources in the example network: independent cyber resources have jurisdiction over physical resources. Cyber-interfaces indicated with red dashed edges. Cyber-physical interfaces indicated with grey dashed edges

Fig. 4.23 Controller adjacency matrix for Example 4.7 [8, 9]

	Q1	Q2	Q3
Q1	1	1	1
Q2	1	1	1
Q3	1	1	1

Legend:

Q1: Water Utility

Q2: Electric Power Utility

Q3: End User

Example 4.7 In Example 4.6, both the cyber-resources and the cyber-interfaces were added to the SysML block diagram, with the cyber-interfaces displayed as associations in blue. Figure 4.22 now expands the original 4-node example network to include the cyber-interfaces as well. From the figures it is clear that all three cyber-resources exchange information. The electric power utility supplies power to both the water treatment facility and the end user. Additionally, the end user consumes water delivered by the water utility. The resulting matrix A_C in Figure 4.23 has size $\sigma(Q) \times \sigma(Q) = 3 \times 3$, with all 9 elements filled. ■

4.5 Service as Operand Behavior

The previous subsections focused on a system's capabilities as sentences that describe what an engineering system does. It identified those capabilities (in Section 4.1), connected them into parallel and serial arrangements (in Section 4.2), and described their control structure (in Sections 4.3 and 4.4). Each of these capabilities

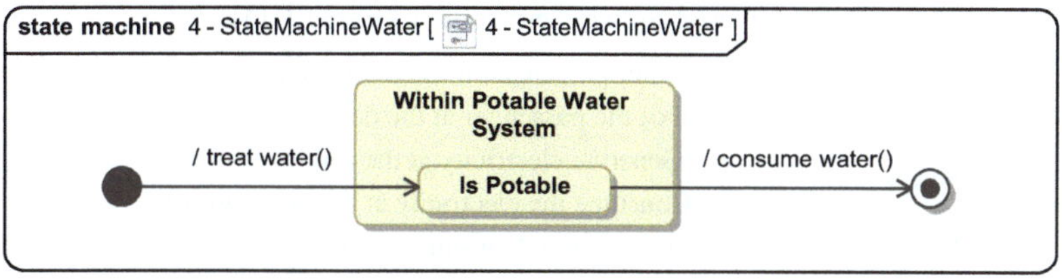

Fig. 4.24 State machine for the service *deliver water*

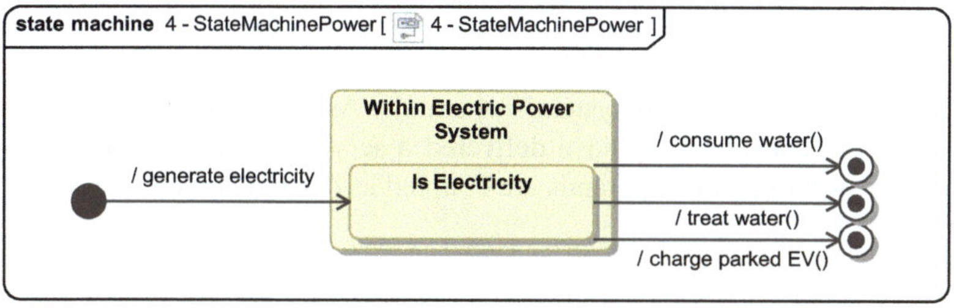

Fig. 4.25 State machine for the service *deliver power*

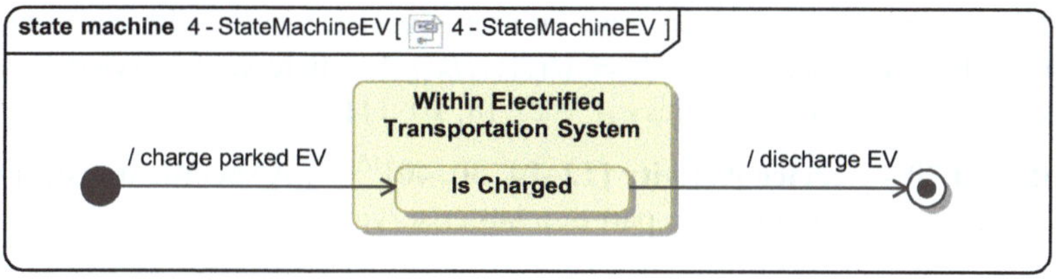

Fig. 4.26 State machine for the service *deliver EV*

acts on a physical operand that when delivered across the engineering system boundary constitutes a service. This section recognizes that as these capabilities act on the operand, they sometimes change its state. Therefore, it is useful in these cases to track these state changes as an operand behavior.

In the SysML modeling language, service delivery is achieved by fulfilling the sequence of processes from input to output in the activity diagram. As soon as the final output is generated, the system service is considered delivered. As mentioned, the state of operands changes as a result of the processes in the activity diagram. In order to describe the behavior of operands in the system, SysML defines a state machine that tracks the state and state transitions of operands or (sub-)systems. The state transitions in the state machine are directly related to the processes in the activity diagram [25, 50]. The state machines for operands *water*, *electric power*, and *electric vehicle* in the 4-node example network are presented in Figures 4.24, 4.25, and 4.26.

In order to maintain ontological lucidity, hetero-functional graph theory recognizes that the structure of the operand is distinct from the structure of the engineering system that acts upon it. Furthermore, the state of the engineering system is also

Table 4.4 Examples of System Services in LFESs [12–14]

Transportation:	{Enter passenger at the origin station, Exit the passenger at the destination}
Power grid:	{Generate electricity at the origin, Consume the electricity at the destination}
Water distribution:	{Treat water, Consume water}
Production:	{Enter the part to an input buffer, Mill the part, Drill a hole in the part, Polish the part, Exit the part from an output buffer}

distinct from the state of the system's operands. As the operand transitions to its final state, it is considered to have delivered a service. Naturally, the state of the engineering system and its operands are coupled as discussed later in Section 4.6.

4.5.1 Service Delivery as Service Net

An engineering system delivers one or more services; each with its associated operand. The set of services is $L = \{l_1, \ldots, l_{\sigma(L)}\}$, where each service l_i has its associated set of service activities $e_{xl_i} \in \mathcal{E}_{l_i}$ [8, 10, 11].

Definition 4.20 – Service Activity [12–14, 30, 40][16,17]: A specific transformation process that may be applied as a part of a larger service. ■

Table 4.4 shows examples of service activities for a broad range of engineering systems.

Relatively simple services may be described as a sequence of service activities [11]:

$$z_{l_i} = e_{x_1 l_i} e_{x_2 l_i} \ldots e_{x_{\sigma(\mathcal{E}_{l_i})} l_i} \tag{4.34}$$

[16]In earlier works where hetero-functional graph theory was applied to production systems [8, 10], services represented manufactured products. The transformation processes necessary to evolve the production of a product were called product events. As the theory found new application domains, however, the more generic term of service was adopted to replace the term products. Furthermore, the term service activity was adopted to align with the SysML language. While the delivery of services and products do have their differences, hetero-functional graph theory assumes that the delivery of a product constitutes a service and hence treats the two concepts equally.

[17]The term "service" within hetero-functional graph theory may be counter-intuitive to some audiences. For example, some might say that services are associated with the execution of the engineering system's capabilities. Such a view, however, is not consistent with the production system literature. Products and services are delivered only after a ***value-adding*** process has happened. It is not sufficient, for example, to simply hold or move an operand within an engineering system. Consequently, it is necessary to track the intermediate states of a product as it gains value prior to delivery across the system boundary.

This definition, however, is insufficient for complex services such as those found in healthcare [29–32] and manufacturing systems [8, 10, 39, 40] where multiple pathways and parallelism are often necessary. Service nets allow service activities to be connected into arbitrary parallel and serial arrangements.

Definition 4.21 – Service Petri Net [12–14, 30, 40][18]: Given service l_i, a service net is described as a tuple:

$$N_{l_i} = \{S_{l_i}, \mathcal{E}_{l_i}, M_{l_i}, W_{l_i}, Q_{l_i}\} \tag{4.35}$$

where

- N_{l_i} is the service net.
- S_{l_i} is the set of places describing a set of service states.
- $\mathcal{E}_{l_i}$ is the set of transitions describing service activities.
- $M_{l_i} \subseteq (S_{l_i} \times \mathcal{E}_{l_i}) \cup (\mathcal{E}_{l_i} \times S_{l_i})$ is the set of arcs describing the relations of (service states to service activities) and (service activities to service states).
- $W_{l_i} : M_{l_i} \rightarrow [0 \ldots 1]$ is the set of weights on the arcs describing the service transition probabilities for the arcs.
- Q_{l_i} is the Petri net marking representing the set of service states.

■

The service Petri net structure implies the following discrete-event dynamics:

Definition 4.22 – Timed Petri Net (Discrete-Event Dynamics [38]): Given a binary input firing vector $U^+_{l_i}[k]$ and a binary output firing vector $U^-_{l_i}[k]$ both of size $\sigma(\mathcal{E}_{l_i}) \times 1$, and the positive and negative components $M^+_{l_i}$ and $M^-_{l_i}$ of the Petri net incidence matrix of size $\sigma(S_{l_i}) \times \sigma(\mathcal{E}_{l_i})$, the evolution of the marking vector Q_{l_i} is given by the state transition function $\Phi_T(Q_{l_i}[k], U^-_{l_i}[k], U^+_{l_i}[k])$:

$$Q_{l_i}[k+1] = \Phi_T(Q_{l_i}[k], U^-_{l_i}[k], U^+_{l_i}[k]) \tag{4.36}$$

where $Q_{l_i} = [Q_{Sl_i}; Q_{El_i}]$ and

$$Q_{Sl_i}[k+1] = Q_{Sl_i}[k] + M^+_{l_i} U^+_{l_i}[k] - M^-_{l_i} U^-_{l_i}[k] \tag{4.37}$$

$$Q_{El_i}[k+1] = Q_{El_i}[k] - U^+_{l_i}[k] + U^-_{l_i}[k] \tag{4.38}$$

■

Example 4.8 Example 4.2 introduced a simple infrastructure system in a SysML block diagram and a SysML activity diagram with swim lanes. The system

[18] In earlier works where hetero-functional graph theory was applied to production systems [8, 10], service nets were called product nets. In more recent work on healthcare delivery systems, service nets are called health nets and represent the health of an individual (patient).

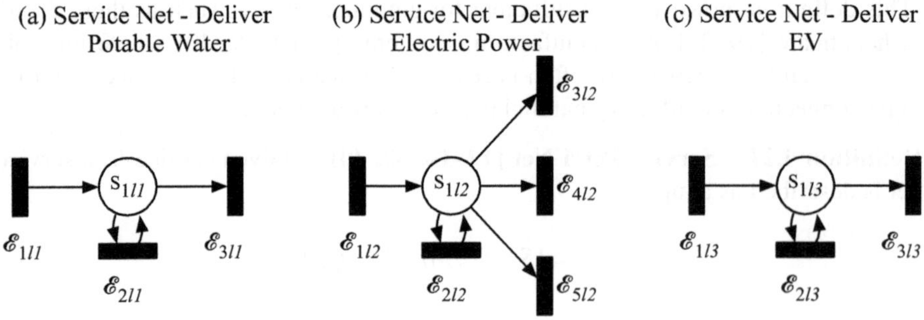

Fig. 4.27 Service nets: three service nets in the 4-node example network. Operands from left to right: (**a**) water, (**b**) power, and (**c**) electric vehicle

delivers three services: deliver potable water, deliver electric power, and deliver EV. These services were introduced in Figures 4.24, 4.25, and 4.26 as SysML state machine diagrams. Hetero-functional graph theory uses service nets (as defined in Definition 4.21) to describe the service delivery behavior. In order to define the service net, each state of the operands is assigned a place. A transition is assigned to the service activities that evolve the state of the operand. The arcs between the places and transitions are based on the inputs and outputs of each of the transitions. Furthermore, an additional "maintain operand state" transition is connected with two one-way arcs for each of the places in the service net.[19]

The service *deliver water* can, therefore, be described using the service net in Figure 4.27a. The service net has one place: {Is Potable}, and three transitions: {treat water(), maintain potable water(), consume water()}. The first inputs the operand into the system, the second maintains its existence until the third outputs the operand from the system. The non-potable water states are not of concern in the 4-node example network and are considered outside the system boundary.

The service *deliver electricity* is described as a service net in Figure 4.27b. The service net has one place: {Is Electricity}, and five transitions: {generate electricity(), maintain electric power(), treat water(), consume water(), charge parked EV()}. These five transitions highlight the interdependent nature of a city's infrastructure systems, as the delivery of electric power is critical to delivering other services in the city.

The service *deliver electric vehicle* is described as a service net in Figure 4.27c. The service net has one place: {Is Charged}, and three transitions: {charge parked EV(), maintain state-of-charge EV(), discharge EV()}. Note that this is a discretization of the continuous state of charge of an electric vehicle.[20] ■

[19]The addition of a single "maintain operand state" transition for each place is absolutely necessary once holding processes of a transformative nature are added to the model.

[20]For a more in-depth look into continuous Petri nets, the reader is referred to [5].

4.5.2 Service Delivery as Service Graph

The service net introduced above can now be translated to a service graph that defines the adjacency of service activities. The translation of a Petri net into a graph is usually achieved by translating the places into nodes, and the transitions into edges. The dual-adjacency matrix of the translated graph calculates the adjacency of the directed edges, and thus the service activities [8, 10]. The service graph is, therefore, the dual-adjacency matrix of the service net. It is calculated as:

$$A_{l_i} = M_{l_i}^{+T} M_{l_i}^{-} \tag{4.39}$$

where M^+ is the positive incidence matrix of the service net with size $\sigma(S_{l_i}) \times \sigma(\mathcal{E}_{l_i})$, and M^- is the negative incidence matrix of the service net with size $\sigma(S_{l_i}) \times \sigma(\mathcal{E}_{l_i})$. The resulting matrix A_{l_i} has size $\sigma(\mathcal{E}_{l_i}) \times \sigma(\mathcal{E}_{l_i})$, and shows the adjacency of service activities in the service net. This adjacency matrix is then represented as a graph. The transitions are its nodes, and the directed arcs represent their adjacency [8, 10].

Example 4.9 In order to convert the service nets into service graphs, the incidence matrices of each of the service nets serve as an input to Equation 4.39. The dual-adjacency matrices are:

$$A_{l_1} = M_{\text{water}}^{+T} M_{\text{water}}^{-} = \begin{bmatrix} 1 \\ 1 \\ 0 \end{bmatrix} \begin{bmatrix} 0 & 1 & 1 \end{bmatrix} = \begin{bmatrix} 0 & 1 & 1 \\ 0 & 1 & 1 \\ 0 & 0 & 0 \end{bmatrix} \tag{4.40}$$

$$A_{l_2} = M_{\text{power}}^{+T} M_{\text{power}}^{-} = \begin{bmatrix} 1 \\ 1 \\ 0 \\ 0 \\ 0 \end{bmatrix} \begin{bmatrix} 0 & 1 & 1 & 1 & 1 \end{bmatrix} = \begin{bmatrix} 0 & 1 & 1 & 1 & 1 \\ 0 & 1 & 1 & 1 & 1 \\ 0 & 0 & 0 & 0 & 0 \\ 0 & 0 & 0 & 0 & 0 \\ 0 & 0 & 0 & 0 & 0 \end{bmatrix} \tag{4.41}$$

$$A_{l_3} = M_{\text{EV}}^{+T} M_{\text{EV}}^{-} = \begin{bmatrix} 1 \\ 1 \\ 0 \end{bmatrix} \begin{bmatrix} 0 & 1 & 1 \end{bmatrix} = \begin{bmatrix} 0 & 1 & 1 \\ 0 & 1 & 1 \\ 0 & 0 & 0 \end{bmatrix} \tag{4.42}$$

These adjacency matrices are visualized in Figure 4.28. ■

4.6 Service Feasibility Matrix

The service feasibility matrix couples the structure of the engineering system to the structure of its services. As the state of the engineering system's operands evolve, the state of the engineering system itself must also evolve. Therefore, these two structures are inherently coupled.

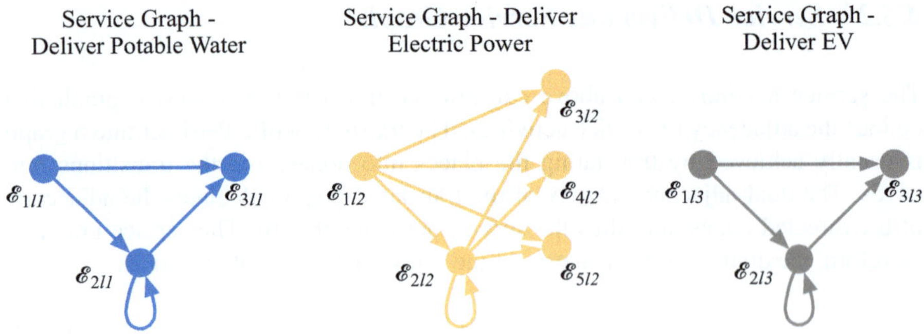

Fig. 4.28 Service graphs: three service graphs in the 4-node example network. Operands from left to right: (**a**) water, (**b**) power, and (c) electric vehicle

In the SysML modeling language, state machines are inherently coupled to activity diagrams. The triggers between states in the state machine correspond to the actions in the activity diagram. The state machines in Figures 4.24, 4.25, and 4.26 can be traced back to the activity diagram in Figure 4.29 (repetition of Figure 4.3). Furthermore, Figure 4.29 shows that the operand states reflected in the state machine diagram are also reflected in the activity diagram as operand or output statements. Additionally, the service model is also integrated in the SysML block diagram as presented in Figure 4.30. The physical resources perform services and are coupled to services in the block diagram by association. Furthermore, services potentially interface with themselves which is represented by self-association [23].

In hetero-functional graph theory, the coupling of the engineering system structure with the operand's behavior as a service is defined by service feasibility matrices [8, 10, 11, 14]. In addition to the coupling of the operand states to the system state, the service feasibility matrix also allows for the calculation of measures of customization and redundancy [8, 19]. This section, first, introduces three feasibility matrices that establish the coupling in Section 4.6.1. It, then, introduces the service selector matrices that allow for the development of service-oriented measures of customization and redundancy in Section 4.6.2.

4.6.1 Service Feasibility Matrix Definitions

By their nature, transformation capabilities, when executed, have the ability to change the operand's state, either in place, or by injecting them across the system boundary. The service transformation feasibility matrix links service transitions to the transformation capabilities that can execute them.

Definition 4.23 – Service Transformation Feasibility Matrix [8, 11, 12, 14]: For a given service l_i, a binary matrix of size $\sigma(\mathcal{E}_{l_i}) \times \sigma(P_\mu)$ whose value $\Lambda_{\mu i}(x, j) = 1$ if e_{xl_i} realizes transformation process $p_{\mu j}$. ∎

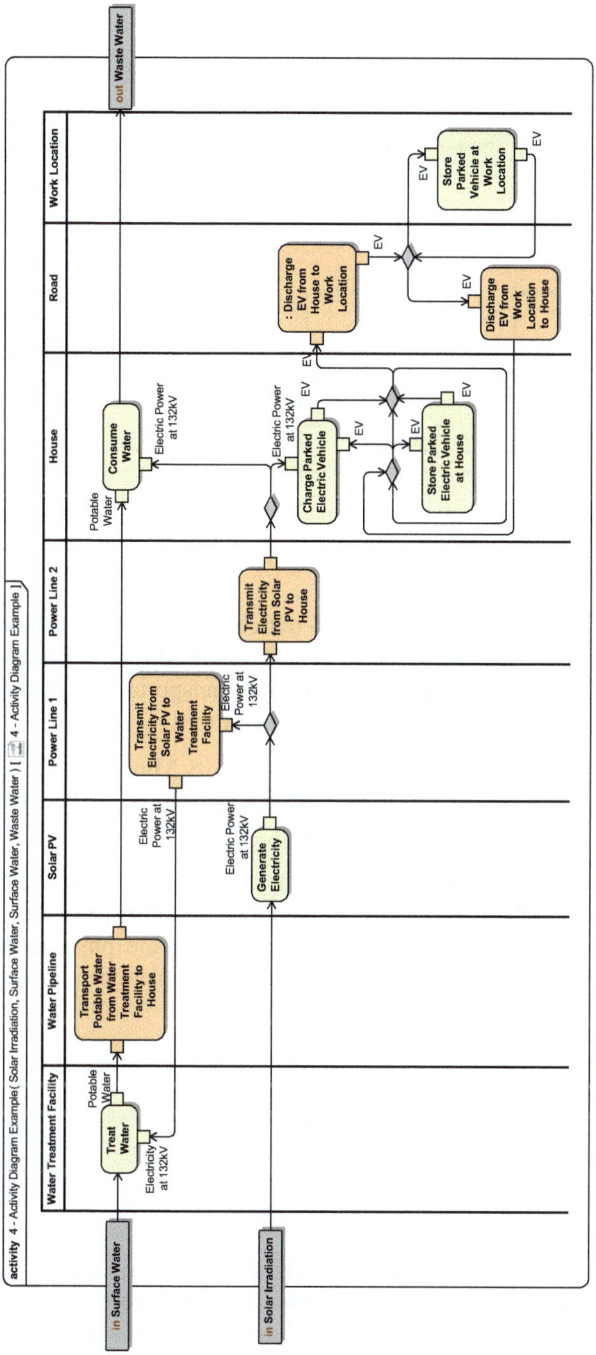

Fig. 4.29 A SysML Activity Diagram: swim lanes allocate function to form for the 4-node smart city network as presented in Figure 4.1. The network consists of transportation, electricity, and water infrastructure

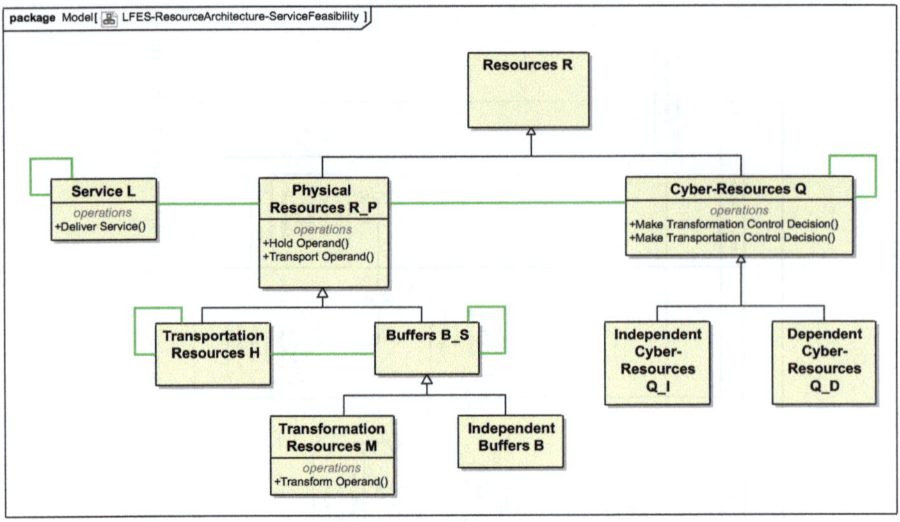

Fig. 4.30 A SysML Block Diagram: the meta-architecture of the system form of an LFES with cyber-resources and the service model

While transportation processes normally do not change the state of their operands, the holding processes of a transformative nature, as introduced in Section 4.1, have the potential to do so. Consequently, the service transportation feasibility matrix couples these holding processes to the service transitions.

Definition 4.24 – Service Transportation Feasibility Matrix [8, 11, 12, 14][21,22]: For a given service l_i, a binary matrix of size $\sigma(\mathcal{E}_{l_i}) \times \sigma(P_\gamma)$ whose value $\Lambda_{\gamma i}(x,g) = 1$ if e_{xl_i} realizes holding process $p_{\gamma g}$. ■

Note that the service transportation feasibility matrix couples the service transitions to the holding processes, rather than the refined transportation processes.

In order for each service activity to be realized by the physical engineering system, there must exist exactly one associated transformation or holding process [8, 10].

$$\sum_{j}^{\sigma(P_\mu)} \Lambda_{\mu i}(x,j) + \sum_{g}^{\sigma(P_\gamma)} \Lambda_{\gamma i}(x,g) = 1 \tag{4.43}$$

[21]Originally, $\Lambda_{\gamma i}$ was defined as a binary matrix of size $1 \times \sigma(P_\gamma)$. However, it has since been expanded to address the potentially transformative nature of holding processes.

[22]The introduction of the "maintain operand state()" transition in the service net mentioned on Page 71 now requires a link to a single holding process of a non-transformative nature so as to indicate that a holding process is associated with the operand that it holds.

Having zero would mean that the service activity would not occur. Having two would mean that the set of transformation and holding processes are not mutually exclusive and an absence of ontological laconicity.

Together, the service transformation and transportation feasibility matrices can be integrated to provide a system-wide sense of the coupling between system processes and service transitions.

Definition 4.25 – Service Feasibility Matrix: For a given service l_i, a binary matrix of size $\sigma(\mathcal{E}_{l_i}) \times \sigma(P)$ whose value $\Lambda_i(x, w) = 1$ if e_{xl_i} realizes process p_w.

$$\Lambda_i = \left[\Lambda_{\mu i} \mid \Lambda_{\gamma i} \otimes \mathbb{1}^{\sigma(P_\eta)T}\right] \tag{4.44}$$

■

Example 4.10 This example couples the service behavior from Example 4.8 to the structural model from Example 4.2. To that end, it uses the service feasibility matrices. First, the *service transformation feasibility matrix* couples the system transformation processes P_μ to the service activities $\mathcal{E}_{l_i}$. Recall that the set of transformation processes consists of: {treat water, generate electricity, consume water}.

Deliver Potable Water: The set of service activities for the operand water $\mathcal{E}_{l_1}$ contains: {treat water, maintain potable water, consume water}. The service transformation feasibility matrix for the operand water $\Lambda_{\mu l_1}$, therefore, has a size of $\sigma(\mathcal{E}_{l_1}) \times \sigma(P_\mu) = 3 \times 3$, with two filled elements:

$$\Lambda_{\mu l_1} = \begin{bmatrix} 1 & 0 & 0 \\ 0 & 0 & 0 \\ 0 & 0 & 1 \end{bmatrix} \tag{4.45}$$

Deliver Electricity: The set of service activities for the operand electricity $\mathcal{E}_{l_2}$ contains: {generate electricity, maintain electric power, treat water, consume water, charge parked EV}. The service transformation feasibility matrix for the operand water $\Lambda_{\mu l_2}$, therefore, has a size of $\sigma(\mathcal{E}_{l_2}) \times \sigma(P_\mu) = 5 \times 3$, with three filled elements:

$$\Lambda_{\mu l_2} = \begin{bmatrix} 0 & 1 & 0 \\ 0 & 0 & 0 \\ 1 & 0 & 0 \\ 0 & 0 & 1 \\ 0 & 0 & 0 \end{bmatrix} \tag{4.46}$$

Deliver Electric Vehicle: The set of service activities for the operand EV $\mathcal{E}_{l_3}$ contains: {charge parked EV, maintain state-of-charge EV, discharge EV}. The service transformation feasibility matrix for the operand EV $\Lambda_{\mu l_3}$, therefore, has size $\sigma(\mathcal{E}_{l_3}) \times \sigma(P_\mu) = 3 \times 3$, with zero filled elements. None of the activities are realized by transformation processes.

$$\Lambda_{\mu l_3} = \begin{bmatrix} 0\,0\,0 \\ 0\,0\,0 \\ 0\,0\,0 \end{bmatrix} \tag{4.47}$$

Second, the *service transportation feasibility matrix* couples the system holding processes P_γ to the service activities $\mathcal{E}_{l_i}$. Recall that the set of holding processes consists of: {carry potable water, carry electricity, charge electric vehicle, discharge electric vehicle, carry electric vehicle}.

Deliver Potable Water: The service transportation feasibility matrix for the operand water $\Lambda_{\gamma l_1}$ has size $\sigma(\mathcal{E}_{l_1}) \times \sigma(P_\gamma) = 3 \times 5$. The matrix contains one filled element that maps the holding process *carry water* to the service activity *maintain potable water*:

$$\Lambda_{\gamma l_1} = \begin{bmatrix} 0\,0\,0\,0\,0 \\ 1\,0\,0\,0\,0 \\ 0\,0\,0\,0\,0 \end{bmatrix} \tag{4.48}$$

Deliver Electricity: The service transportation feasibility matrix for the operand electricity $\Lambda_{\gamma l_2}$ has size $\sigma(\mathcal{E}_{l_2}) \times \sigma(P_\gamma) = 5 \times 5$. The matrix contains two filled elements. The first couples the holding process *charge parked EV* to its service activity. The second couples the service activity *maintain electric power* to the holding process *carry electricity*:

$$\Lambda_{\gamma l_2} = \begin{bmatrix} 0\,0\,0\,0\,0 \\ 0\,1\,0\,0\,0 \\ 0\,0\,0\,0\,0 \\ 0\,0\,0\,0\,0 \\ 0\,0\,1\,0\,0 \end{bmatrix} \tag{4.49}$$

Deliver Electric Vehicle: The service transportation feasibility matrix for the operand EV $\Lambda_{\gamma l_3}$ has size $\sigma(\mathcal{E}_{l_3}) \times \sigma(P_\gamma) = 3 \times 5$. The matrix contains three filled elements. The first two elements couple the holding processes *charge parked EV* and *discharge EV* to the service activities. The third element couples *park EV* to the service activity *maintain state-of-charge EV*:

$$\Lambda_{\gamma l_3} = \begin{bmatrix} 0\,0\,1\,0\,0 \\ 0\,0\,0\,0\,1 \\ 0\,0\,0\,1\,0 \end{bmatrix} \tag{4.50}$$

Finally, the *service feasibility matrices* are calculated based on the two previous matrices, using Equation 4.44:

- The service feasibility matrix for the service *deliver potable water* is calculated as follows:

$$\Lambda_1 = \left[\Lambda_{\mu 1} \mid \Lambda_{\gamma 1} \otimes \mathbb{1}^{\sigma(P_\eta)T}\right] \tag{4.51}$$

where Λ_1 has size $\sigma(\mathcal{E}_{l_1}) \times \sigma(P) = 3 \times 83$, with 86 filled elements.
- The service feasibility matrix for the service *deliver electricity* is calculated as follows:

$$\Lambda_2 = \left[\Lambda_{\mu 2} \mid \Lambda_{\gamma 2} \otimes \mathbb{1}^{\sigma(P_\eta)T}\right] \tag{4.52}$$

where Λ_2 has size $\sigma(\mathcal{E}_{l_2}) \times \sigma(P) = 5 \times 83$, with 163 filled elements.
- The service feasibility matrix for the service *deliver electric vehicle* is calculated as follows:

$$\Lambda_3 = \left[\Lambda_{\mu 3} \mid \Lambda_{\gamma 3} \otimes \mathbb{1}^{\sigma(P_\eta)T}\right] \tag{4.53}$$

where Λ_3 has size $\sigma(\mathcal{E}_{l_3}) \times \sigma(P) = 3 \times 83$, with 240 filled elements.

■

4.6.2 Service Degrees of Freedom

The service feasibility matrices described above are now used to calculate how many system capabilities apply to a service activity, to a service as a whole, or the full line (or set) of services. When enumerated these are called service degrees of freedom. Table 4.5 provides service selector matrices that are essential to the calculation of the service degrees of freedom.

Definition 4.26 – Service Transformation Degrees of Freedom [10, 11, 14]: The set of independent service events $\mathcal{E}_{LM}$ that completely define the available combinations of transformation process and resource that are required by the delivery of the service. ■

Definition 4.27 – Service Transportation Degrees of Freedom [10, 11, 14]: The set of independent service events $\mathcal{E}_{LH}$ that completely define the available combinations of transportation process and resource that can be utilized by the delivery of the service. ■

Definition 4.28 – Service Degrees of Freedom [10, 11, 14]: The set of independent service events $\mathcal{E}_{LS}$ that completely define the available combinations of transformation or transportation process and resource that can be utilized by the delivery of the service. ■

Table 4.5 Types of service selector matrices [10, 11, 14]

	Symbol	Formula	Scope
Transformation	Λ_{Mxi}	$\left[e_x^T \Lambda_{\mu i}\right]^T \mathbb{1}^{\sigma(M)T}$	Service activity
	Λ_{Mi}	$\left[\mathbb{1}^{\sigma(\mathcal{E}_{l_i})T} \odot \Lambda_{\mu i}\right]^T \mathbb{1}^{\sigma(M)T}$	Service
	Λ_{ML}	$\bigvee_i^{\sigma(L)} \Lambda_{Mi}$	Service line
Transportation	Λ_{Hxi}	$\left[\left[e_x^T \Lambda_{\gamma i}\right] \otimes \mathbb{1}^{\sigma(P_\eta)T}\right]^T \mathbb{1}^{\sigma(R)T}$	Service
	Λ_{Hi}	$\left[\mathbb{1}^{\sigma(\mathcal{E}_{l_i})T} \odot \Lambda_{\gamma i} \otimes \mathbb{1}^{\sigma(P_\eta)T}\right]^T \mathbb{1}^{\sigma(R)T}$	Service
	Λ_{HL}	$\bigvee_i^{\sigma(L)} \Lambda_{Hi}$	Service line
Transformation and transportation	Λ_{Sxi}	$\left[e_x^T \Lambda_i\right]^T \mathbb{1}^{\sigma(R)T}$	Service activity
	Λ_{SMxi}	$\left[\begin{array}{c\|c} \Lambda_{Mxi} & \mathbf{1} \\ \hline \multicolumn{2}{c}{\mathbf{0}} \end{array}\right]$	Service activity
	Λ_{SHxi}	$\left[\begin{array}{c\|c} \mathbf{0} & \mathbf{0} \\ \hline \multicolumn{2}{c}{\Lambda_{Hxi}} \end{array}\right]$	Service activity
	Λ_{Si}	$\left[\begin{array}{c\|c} \Lambda_{Mi} & \mathbf{1} \\ \hline \multicolumn{2}{c}{\Lambda_{Hi}} \end{array}\right] = \left[\mathbb{1}^{\sigma(\mathcal{E}_{l_i})T} \odot \Lambda_i\right]^T \mathbb{1}^{\sigma(R)T}$	Service
	Λ_{SHi}	$\left[\begin{array}{c\|c} \mathbf{0} & \mathbf{0} \\ \hline \multicolumn{2}{c}{\Lambda_{Hi}} \end{array}\right]$	Service
	Λ_{SL}	$\left[\begin{array}{c\|c} \Lambda_{ML} & \mathbf{1} \\ \hline \multicolumn{2}{c}{\Lambda_{HL}} \end{array}\right] = \bigvee_i^{\sigma(L)} \Lambda_{Si}$	Service line

In order to complete the calculation, a number of service selector matrices are introduced as an intermediate step. They are all of equal size to the corresponding knowledge base. Table 4.5 summarizes their calculation. For example, Equations 4.54 through 4.56 calculate the number of transformation and transportation capabilities utilized by the full set of services delivered by the engineering system [10, 11, 14].

$$DOF_{LM} = \langle \Lambda_{ML} \cdot J_M, \bar{K}_M \rangle_F \tag{4.54}$$

$$DOF_{LH} = \langle \Lambda_{HL} \cdot J_{\bar{H}}, \bar{K}_{\bar{H}} \rangle_F \tag{4.55}$$

$$DOF_{LS} = \langle \Lambda_{SL} \cdot J_S, \bar{K}_S \rangle_F \tag{4.56}$$

These values are important to understand how capable the engineering system is relative to the services it needs to deliver [8, 19]. Excessively redundant capabilities can be potentially decommissioned to reduce costs without any degradation in the quality of service.

Example 4.11 This example demonstrates the calculation of all three service degree of freedom measures for the 4-node example network (Example 4.2). It builds off the service feasibility matrices calculated in Example 4.10.

First, the service transformation degrees of freedom DOF_{LM} are calculated using Equation 4.54. The transformation knowledge base J_M was calculated in Example 4.4, and Λ_{ML} can be calculated using the service selector matrices in Table 4.5.

$$\Lambda_{ML} = \bigvee_{i}^{\sigma(L)} \Lambda_{Mi} \tag{4.57}$$

where:

$$\Lambda_{Mi} = \left[\mathbb{1}^{\sigma(\mathcal{E}_{l_i})T} \odot \Lambda_{\mu i}\right]^T \mathbb{1}^{\sigma(M)T} \tag{4.58}$$

$$\Lambda_{M1} = \left[\mathbb{1}^{\sigma(\mathcal{E}_{l_1})T} \odot \Lambda_{\mu 1}\right]^T \mathbb{1}^{\sigma(M)T} = \left[[1\ 1\ 1] \odot \begin{bmatrix} 1\,0\,0 \\ 0\,0\,0 \\ 0\,0\,1 \end{bmatrix}\right]^T [1\ 1\ 1] = \begin{bmatrix} 1\ 1\ 1 \\ 0\,0\,0 \\ 1\ 1\ 1 \end{bmatrix} \tag{4.59}$$

$$\Lambda_{M2} = \left[\mathbb{1}^{\sigma(\mathcal{E}_{l_2})T} \odot \Lambda_{\mu 2}\right]^T \mathbb{1}^{\sigma(M)T} = \left[[1\ 1\ 1\ 1\ 1] \odot \begin{bmatrix} 0\,1\,0 \\ 0\,0\,0 \\ 1\,0\,0 \\ 0\,0\,1 \\ 0\,0\,0 \end{bmatrix}\right]^T [1\ 1\ 1] = \begin{bmatrix} 1\ 1\ 1 \\ 1\ 1\ 1 \\ 1\ 1\ 1 \end{bmatrix} \tag{4.60}$$

$$\Lambda_{M3} = \left[\mathbb{1}^{\sigma(\mathcal{E}_{l_3})T} \odot \Lambda_{\mu 3}\right]^T \mathbb{1}^{\sigma(M)T} = \left[[1\ 1\ 1] \odot \begin{bmatrix} 0\,0\,0 \\ 0\,0\,0 \\ 0\,0\,0 \end{bmatrix}\right]^T [1\ 1\ 1] = \begin{bmatrix} 0\,0\,0 \\ 0\,0\,0 \\ 0\,0\,0 \end{bmatrix} \tag{4.61}$$

Therefore:

$$\Lambda_{ML} = \begin{bmatrix} 1\ 1\ 1 \\ 1\ 1\ 1 \\ 1\ 1\ 1 \end{bmatrix} \tag{4.62}$$

Now, the service transformation degrees of freedom are calculated based on J_S, Λ_{ML}, and the transformation constraints matrix K_M (all-zeroes, 3×3):

$$DOF_{LM} = \langle \Lambda_{ML} \cdot J_M, \bar{K}_M \rangle_F = \left\langle \begin{bmatrix} 1 & 1 & 1 \\ 1 & 1 & 1 \\ 1 & 1 & 1 \end{bmatrix} \cdot \begin{bmatrix} 1 & 0 & 0 \\ 0 & 1 & 0 \\ 0 & 0 & 1 \end{bmatrix}, \begin{bmatrix} 1 & 1 & 1 \\ 1 & 1 & 1 \\ 1 & 1 & 1 \end{bmatrix} \right\rangle_F = 3 \tag{4.63}$$

In conclusion, the number of service transformation degrees of freedom is three.

The number of service transportation degrees of freedom is calculated with Equation 4.55. Matrix Λ_{HL} has size 80×8 with all filled elements. Consequently, the number of service transportation degrees of freedom is eight.

Given the calculations for the service transformation and transportation degrees of freedom, the service degrees of freedom follow straightforwardly:

$$DOF_{LS} = \langle \Lambda_{SL} \cdot J_S, \bar{K}_S \rangle_F \tag{4.64}$$

where

$$\Lambda_{SL} = \left[\begin{array}{c|c} \Lambda_{ML} & \mathbf{1} \\ \hline \multicolumn{2}{c}{\Lambda_{HL}} \end{array} \right] \tag{4.65}$$

The total number of service degrees of freedom is, therefore, 11. ■

4.7 The System Adjacency Matrix: An Integrated View of Hetero-functional Graph Theory

The previous sections introduced the fundamental mathematical models in hetero-functional graph theory. This section integrates these models into a single hetero-functional graph represented as a System Adjacency Matrix $\mathbb{A}$. The System Adjacency Matrix provides a holistic representation of the cyber-physical engineering system. As shown in Equation 4.66, the matrix is organized in a matrix block form derived from the six mathematical models discussed previously.

$$\mathbb{A} = \begin{bmatrix} \mathbb{A}_L & \mathbb{A}_{L\rho} & \mathbf{0} \\ \mathbb{A}_{\rho L} & A_\rho & \mathbb{A}_{\rho C} \\ \mathbf{0} & \mathbb{A}_{C\rho} & A_C \end{bmatrix} \tag{4.66}$$

Alternatively, a projected system adjacency matrix can be calculated so as to significantly eliminate sparsity.

$$\widetilde{\mathbb{A}} = \begin{bmatrix} \mathbb{A}_L & \widetilde{\mathbb{A}}_{L\rho} & \mathbf{0} \\ \widetilde{\mathbb{A}}_{\rho L} & \widetilde{A}_\rho & \widetilde{\mathbb{A}}_{\rho C} \\ \mathbf{0} & \widetilde{\mathbb{A}}_{C\rho} & A_C \end{bmatrix} \tag{4.67}$$

This section discusses the unprojected and projected forms of each of these block matrix forms in detail.

In essence, a hetero-functional graph represents an interconnected model of an engineering system including its capabilities, its service model, and its control model. Equation 4.66 shows that the central matrix block of the system adjacency matrix $\mathbb{A}$ is the hetero-functional adjacency matrix A_ρ (discussed in Section 4.2 on Page 55). It represents the logical connections between the system's capabilities. Equation 4.66 also shows that the bottom right matrix block of $\mathbb{A}$ is controller adjacency matrix, A_C (discussed in Section 4.4 on Page 64). It represents the informatic connections between the controller agents in the engineering system. The upper left matrix block $\mathbb{A}_L$ represents the *collection* of service models in the engineering system (discussed in Section 4.5 on Page 68). It includes the connections between the service activities in each of the service models (Petri nets). Consequently, the block matrices $\mathbb{A}_{L\rho}$ and $\mathbb{A}_{\rho L}$ represent the logical coupling of service activities to the engineering system's capabilities (i.e. structural degrees of freedom). They are derived from the service feasibility matrices (discussed in Section 4.6 on Page 73). Similarly, the block matrices $\mathbb{A}_{\rho C}$ and $\mathbb{A}_{C\rho}$ represent the logical coupling of controller agents to the engineering system's capabilities. They are derived from the controller agency matrix (discussed in Section 4.3 on Page 60). Each of these block matrices is now discussed in the order presented above.

Block A_ρ: The hetero-functional adjacency matrix A_ρ is the core of the System Adjacency Matrix $\mathbb{A}$. A_ρ constitutes the first two models of hetero-functional graph theory, as System Concept A_S is necessary for its calculation. It represents the structure of the physical engineering system in terms of structural degrees of freedom as nodes and system-sequence degrees of freedom as edges. It has a size of $\sigma(R_P)\sigma(P) \times \sigma(R_P)\sigma(P)$. The projected hetero-functional adjacency matrix $\widetilde{A}_\rho$ is a projection of A_ρ, as calculated in Equation 4.31. Its size is $\sigma(\mathcal{E}_{l_i}) \times \sigma(\mathcal{E}_{l_i})$.

Block A_C: The controller adjacency matrix A_C is the bottom right block of $\mathbb{A}$. It constitutes the third model of hetero-functional graph theory. It represents the interfaces between the cyber-resources. It has a size of $\sigma(Q) \times \sigma(Q)$.

Block $\mathbb{A}_L$: $\mathbb{A}_L$ is the upper left matrix block of $\mathbb{A}$. As it represents the collection of service models in the engineering system, it constitutes the fourth model of hetero-functional graph theory. It takes service activities as nodes and shows the logical coupling between them. The block diagonal features terms of the form $\mathcal{M}_{li}^{+T}\mathcal{M}_{li}^{-}$, as demonstrated in Equation 4.68. Recall that the positive and negative components of the service Petri net incidence matrix for a given service $\mathcal{M}_{li}^{+}$ and $\mathcal{M}_{li}^{-}$ create a bipartite graph between service activities and places. The product $\mathcal{M}_{li}^{+T}\mathcal{M}_{li}^{-}$ creates an adjacency matrix between the service activities. The block diagonal form shows that activities in one service are only coupled within the same service. Distinct services are entirely uncoupled.

$$\mathbb{A}_L = \begin{bmatrix} \mathcal{M}_{l1}^{+T}\mathcal{M}_{l1}^{-} & 0 & \dots & 0 \\ 0 & \mathcal{M}_{l2}^{+T}\mathcal{M}_{l2}^{-} & & \vdots \\ \vdots & & \ddots & 0 \\ 0 & \dots & 0 & \mathcal{M}_{\sigma(L)}^{+T}\mathcal{M}_{\sigma(L)}^{-} \end{bmatrix} \tag{4.68}$$

where $\mathbb{A}_L$ has size $\sum_i \sigma(\mathcal{E}_{l_i}) \times \sum_i \sigma(\mathcal{E}_{l_i})$.

Block $\mathbb{A}_{\rho C}$ and $\mathbb{A}_{C\rho}$: These two block matrices couple the structural degrees of freedom in A_ρ and the independent cyber-resources Q. In this regard, the controller agency matrix as the fifth model of hetero-functional graph theory proves useful. Rather than using A_Q which couples the set of resources R to the set of physical resources R_P, it is more useful to use the sub-matrix $\overline{A}_Q$ which only shows the jurisdiction of independent cyber-resources Q over the physical resources R_P. The calculation of $\mathbb{A}_{\rho C}$ then follows straightforwardly:

$$\mathbb{A}_{\rho C} = \overline{A}_Q \otimes \mathbb{1}^{\sigma(P)} \tag{4.69}$$

where the term $\mathbb{1}^{\sigma(P)}$ is introduced to create a coupling to all the processes that pertain to the physical resources R_P. The couplings captured in $\mathbb{A}_{\rho C}$ represents the *sensed* information coming from the structural degrees of freedom to the cyber-resources, while the couplings captured in $\mathbb{A}_{C\rho}$ represent the *actuation* signals sent in the opposite direction. Consequently, the size of $\mathbb{A}_{\rho C}$ is $\sigma(R_P)\sigma(P) \times \sigma(Q)$ and the size of $\mathbb{A}_{C\rho}$ is its transpose. Their projected form is calculated as follows:

$$\widetilde{\mathbb{A}}_{\rho C} = \mathbb{P}_S \mathbb{A}_{\rho C} \tag{4.70}$$

$$\widetilde{\mathbb{A}}_{C\rho} = \mathbb{A}_{C\rho} \mathbb{P}_S^T \tag{4.71}$$

Blocks $\mathbb{A}_{L\rho}$ and $\mathbb{A}_{\rho L}$: These two block matrices couple the transitions in the service Petri nets $\mathcal{E}_{l_i}$ to the structural degrees of freedom in A_ρ. In this regard, the service feasibility matrices as the sixth (and last) model of hetero-functional graph theory proves useful. The service feasibility matrix $\mathbb{A}_{L\rho}$ is calculated as:

$$\mathbb{A}_{L\rho} = \mathbb{A}_{\rho L}^T = \begin{bmatrix} \widehat{\Lambda}_{l_1} \\ \widehat{\Lambda}_{l_2} \\ \vdots \\ \widehat{\Lambda}_{\sigma(L)} \end{bmatrix} \tag{4.72}$$

where

$$\widehat{\Lambda}_{l_i} = \mathbb{1}^{\sigma(R_P)T} \otimes \Lambda_{l_i} \tag{4.73}$$

which has a size of $\sigma(\mathcal{E}_{l_i}) \times \sigma(R_P)\sigma(P)$. The size of matrix $\mathbb{A}_{L\rho}$ is consequently $\sum_i \sigma(\mathcal{E}_{l_i}) \times \sigma(R_P)\sigma(P)$. Their projected form is calculated as follows:

$$\widetilde{\mathbb{A}}_{\rho L} = \mathbb{P}_S \mathbb{A}_{\rho L} \tag{4.74}$$

$$\widetilde{\mathbb{A}}_{L\rho} = \mathbb{A}_{L\rho} \mathbb{P}_S^T \tag{4.75}$$

Example 4.12 This example integrates all examples in the chapter related to the 4-node example network to create the system adjacency matrix. Given the size of the matrices, even for a small 4-node network, the example constructs the projected system adjacency matrix. As a conclusion, the projected system adjacency matrix is visualized in Figure 4.31.

The projected system adjacency matrix consists of seven matrix blocks.

- The *first block*, the projected hetero-functional adjacency matrix $\widetilde{A}_\rho$ is directly drawn from Example 4.5:

$$\widetilde{A}_\rho = \begin{bmatrix} 0&0&0&0&0&0&1&0&0&0&0 \\ 0&0&0&0&0&0&0&1&1&0&0 \\ 0&0&0&0&0&0&0&0&0&0&0 \\ 0&0&0&1&1&0&0&0&0&1&0 \\ 0&0&0&1&1&0&0&0&0&1&0 \\ 0&0&0&0&0&1&0&0&0&0&1 \\ 0&0&1&0&0&0&0&0&0&0&0 \\ 1&0&0&0&0&0&0&0&0&0&0 \\ 0&0&1&1&0&0&0&0&0&0&0 \\ 0&0&0&0&0&1&0&0&0&0&1 \\ 0&0&0&1&1&0&0&0&0&1&0 \end{bmatrix} \tag{4.76}$$

- The *second block*, the controller adjacency matrix A_C is directly drawn from Example 4.7:

$$A_C = \begin{bmatrix} 1&1&1 \\ 1&1&1 \\ 1&1&1 \end{bmatrix} \tag{4.77}$$

- The *third block*, matrix $\mathbb{A}_L$ is constructed using the three service adjacency matrices from Example 4.9:

$$\mathbb{A}_L = \begin{bmatrix} A_{l_1} & \mathbf{0} & \mathbf{0} \\ \mathbf{0} & A_{l_2} & \mathbf{0} \\ \mathbf{0} & \mathbf{0} & A_{l_3} \end{bmatrix} = \begin{bmatrix} \begin{bmatrix} 0&1&1 \\ 0&1&1 \\ 0&0&0 \end{bmatrix} & \mathbf{0} & \mathbf{0} \\ \mathbf{0} & \begin{bmatrix} 0&1&1&1&1 \\ 0&1&1&1&1 \\ 0&0&0&0&0 \\ 0&0&0&0&0 \\ 0&0&0&0&0 \end{bmatrix} & \mathbf{0} \\ \mathbf{0} & \mathbf{0} & \begin{bmatrix} 0&1&1 \\ 0&1&1 \\ 0&0&0 \end{bmatrix} \end{bmatrix} \tag{4.78}$$

- The *fourth and fifth* blocks are matrices $\widetilde{\mathbb{A}}_{\rho C}$ and $\widetilde{\mathbb{A}}_{C\rho}$. These are calculated using the controller agency matrix ($\overline{A}_Q$) from Example 4.6:

$$\widetilde{\mathbb{A}}_{\rho C} = \mathbb{P}_S\left[\overline{A}_Q \otimes \mathbb{1}^{\sigma(P_\eta)}\right] = \begin{bmatrix} 1&0&0 \\ 0&1&0 \\ 0&0&1 \\ 0&0&1 \\ 0&0&1 \\ 0&0&1 \\ 1&0&0 \\ 0&1&0 \\ 0&1&0 \\ 0&0&1 \\ 0&0&1 \end{bmatrix} \tag{4.79}$$

- The *sixth and seventh* blocks are matrices $\widetilde{\Lambda}_{L\rho}$ and $\widetilde{\Lambda}_{\rho L}$. These are calculated as follows:

$$\widetilde{\mathbb{A}}_{L\rho} = \widetilde{\mathbb{A}}_{\rho L}^T = \begin{bmatrix} \mathbb{1}^{\sigma(R)T} \otimes \Lambda_{l_1} \\ \mathbb{1}^{\sigma(R)T} \otimes \Lambda_{l_2} \\ \mathbb{1}^{\sigma(R)T} \otimes \Lambda_{l_3} \end{bmatrix} \mathbb{P}_S^T = \left[\begin{array}{ccccccccccc} 1&0&0&0&0&0&0&0&0&0&0 \\ 0&0&0&0&0&0&1&0&0&0&0 \\ 0&0&1&0&0&0&0&0&0&0&0 \\ \hline 0&1&0&0&0&0&0&0&0&0&0 \\ 0&0&0&0&0&0&0&1&1&0&0 \\ 1&0&0&0&0&0&0&0&0&0&0 \\ 0&0&1&0&0&0&0&0&0&0&0 \\ 0&0&0&1&0&0&0&0&0&0&0 \\ \hline 0&0&0&1&0&0&0&0&0&0&0 \\ 0&0&0&0&1&1&0&0&0&0&0 \\ 0&0&0&0&0&0&0&0&0&1&1 \end{array}\right] \tag{4.80}$$

After the derivation of each of the seven block matrices in the system adjacency matrix, the full system adjacency matrix is calculated using Equation 4.67:

$$
\mathbb{A} = \begin{bmatrix}
\begin{bmatrix}
\begin{matrix}0&1&1\\0&1&1\\0&0&0\end{matrix} & \mathbf{0} & \mathbf{0} \\
\mathbf{0} & \begin{matrix}0&1&1&1&1\\0&1&1&1&1\\0&0&0&0&0\\0&0&0&0&0\\0&0&0&0&0\end{matrix} & \mathbf{0} \\
\mathbf{0} & \mathbf{0} & \begin{matrix}0&1&1\\0&1&1\\0&0&0\end{matrix}
\end{bmatrix} &
\left[\begin{array}{ccccccccccc}
1&0&0&0&0&0&0&0&0&0&0\\
0&0&0&0&0&0&1&0&0&0&0\\
0&0&1&0&0&0&0&0&0&0&0\\ \hline
0&1&0&0&0&0&0&0&0&0&0\\
0&0&0&0&0&0&0&1&1&0&0\\
1&0&0&0&0&0&0&0&0&0&0\\
0&0&1&0&0&0&0&0&0&0&0\\
0&0&0&1&0&0&0&0&0&0&0\\ \hline
0&0&0&1&0&0&0&0&0&0&0\\
0&0&0&0&1&1&0&0&0&0&0\\
0&0&0&0&0&0&0&0&0&1&1
\end{array}\right] & \mathbf{0} \\
\left[\begin{array}{ccc|ccccc|ccc}
1&0&0&0&0&1&0&0&0&0&0\\
0&0&0&1&0&0&0&0&0&0&0\\
0&0&1&0&0&0&1&0&0&0&0\\
0&0&0&0&0&0&0&1&1&0&0\\
0&0&0&0&0&0&0&0&0&1&0\\
0&0&0&0&0&0&0&0&0&1&0\\
0&1&0&0&0&0&0&0&0&0&0\\
0&0&0&0&1&0&0&0&0&0&0\\
0&0&0&0&1&0&0&0&0&0&0\\
0&0&0&0&0&0&0&0&0&0&1\\
0&0&0&0&0&0&0&0&0&0&1
\end{array}\right] &
\begin{bmatrix}
0&0&0&0&0&0&1&0&0&0&0\\
0&0&0&0&0&0&0&1&1&0&0\\
0&0&0&0&0&0&0&0&0&0&0\\
0&0&0&1&1&0&0&0&0&1&0\\
0&0&0&1&1&0&0&0&0&1&0\\
0&0&0&0&0&1&0&0&0&0&1\\
0&0&1&0&0&0&0&0&0&0&0\\
1&0&0&0&0&0&0&0&0&0&0\\
0&0&1&1&0&0&0&0&0&0&0\\
0&0&0&0&0&1&0&0&0&0&1\\
0&0&0&1&1&0&0&0&0&1&0
\end{bmatrix} &
\begin{bmatrix}
1&0&0\\0&1&0\\0&0&1\\0&0&1\\0&0&1\\0&0&1\\1&0&0\\0&1&0\\0&1&0\\0&0&1\\0&0&1
\end{bmatrix} \\
\mathbf{0} &
\begin{bmatrix}
1&0&0&0&0&0&1&0&0&0&0\\
0&1&0&0&0&0&0&1&1&0&0\\
0&0&1&1&1&1&0&0&0&1&1
\end{bmatrix} &
\begin{bmatrix}1&1&1\\1&1&1\\1&1&1\end{bmatrix}
\end{bmatrix} \tag{4.81}
$$

where the final size of the system adjacency matrix is:

$$
\left[\sum_i^{\sigma(L)} \mathcal{E}_{l_i} + \sigma(R)\sigma(P) + \sigma(Q)\right] \times \left[\sum_i^{\sigma(L)} \mathcal{E}_{l_i} + \sigma(R)\sigma(P) + \sigma(Q)\right] = 25 \times 25 \tag{4.82}
$$

Finally, the system adjacency matrix is visualized in Figure 4.31. In this figure, a distinction is made between three separate parts of the system adjacency matrix to maintain readability. The original 4-node example network is presented in Figure 4.31a. The core of the system adjacency matrix is the hetero-functional adjacency matrix, presented in Figure 4.31b. Figure 4.31c presents the visualization of matrices $\widetilde{\mathbb{A}}_{\rho C}$, $\widetilde{\mathbb{A}}_{C\rho}$, and A_C, mapped to the system capabilities. Finally, Figure 4.31d visualizes the service delivery matrices $\widetilde{\mathbb{A}}_{L\rho}$, $\widetilde{\mathbb{A}}_{\rho L}$, and $\mathbb{A}_L$, mapped to the system capabilities. The system capabilities are shared across all three figures and are the interface when the models are superimposed. ■

The System Adjacency Matrix provides a holistic representation of the structure of a cyber-physical engineering system. It, explicitly, addresses both system

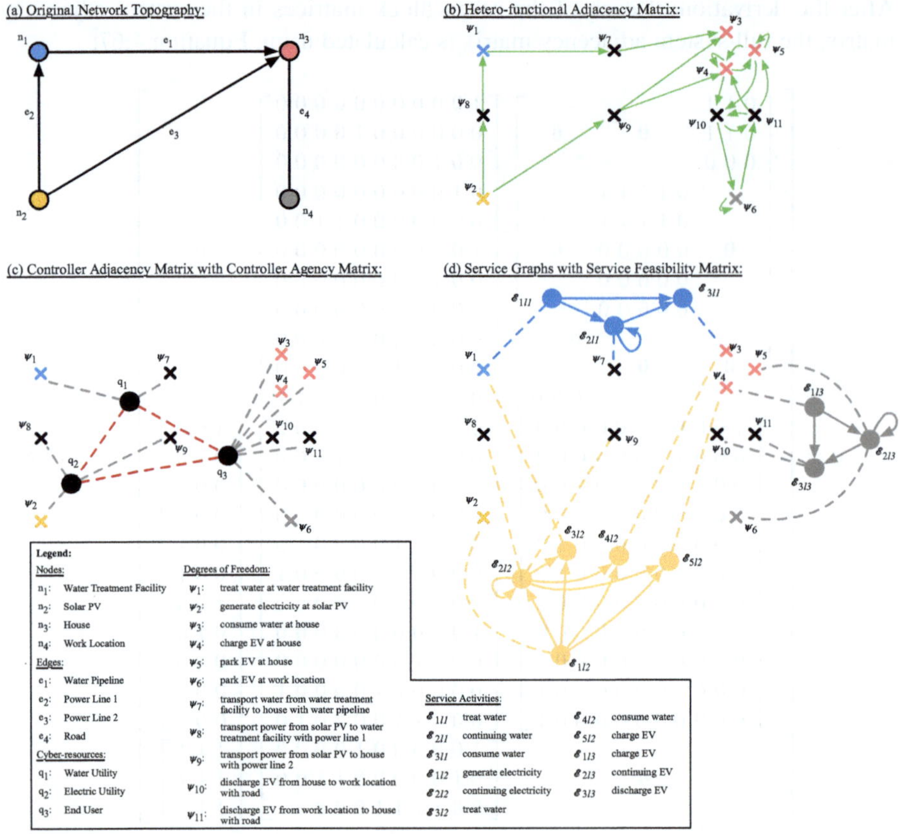

Fig. 4.31 System Adjacency Matrix: a comparison of the original 4-node example network in (**a**) with the hetero-functional adjacency matrix in (**b**), the controller model coupled to the capabilities in (**c**), and the service model coupled to the capabilities in (**d**). Graphs (**b**), (**c**), and (**d**) are three distinct representations of subsets of the system adjacency matrix

function and system form. At its heart, the hetero-functional adjacency matrix A_ρ represents the edges between structural degrees of freedom (i.e., capabilities) as nodes. Similarly, the nodes of the controller adjacency matrix A_C are the controllers Q; each of which is only able to execute a single control or decision algorithm. Finally, the nodes of the $\mathbb{A}_L$ matrix represent the service Petri transitions of each service. In all of these cases, system function and system form are inextricably tied. In this regard, hetero-functional graph theory offers greater ontological lucidity than a graph theory based exclusively on form or function.

4.8 Conclusion

In conclusion, hetero-functional graph theory differentiates itself in that it centers itself on the allocated architecture as it addresses system function and system form simultaneously. This implicitly requires mutually exclusive and collectively exhaustive sets of system processes and resources. It also quantifies a system's capabilities in terms of structural degrees of freedom. The hetero-functional adjacency matrix then couples these capabilities into a network that respects both functional interactions and physical interfaces. ***Because there is an explicit differentiation of system processes, hetero-functional graph theory provides a basis upon which networks with unlike function can be combined into a single mathematical model of system structure.*** Hetero-functional graph theory is also explicitly cyber-physical in that it differentiates between systems of equivalent physical capability but different control structure. The controller adjacency matrix admits a spectrum of control architectures from entirely centralized to entirely distributed. Finally, hetero-functional graph theory recognizes that the execution of the engineering system's capabilities can change the state of the associated operands. The service Petri net describes this state evolution and the service feasibility matrices couples it back to the capabilities of the engineering system. These six matrices describe entirely different phenomena in an engineering system. Their systematic integration results in the system adjacency matrix as a block diagonal form that addresses the full structure of a cyber-physical engineering system. At the beginning of this chapter, Table 4.1 provided an overview of its discussion. Table 4.6 now provides an overview of the elements and the mathematical notation of hetero-functional graph theory.

Given the articulation of these seven graph theoretical models in hetero-functional graph theory, the discussion can return to the motivational example in Chapter 2. Each of these seven matrices has been calculated and worked in Appendix A. Whereas "multi-layer networks" all exhibited a constraint that prevented a complete mathematical description of this motivational example, hetero-functional graph theory straightforwardly provides a mathematical model. Furthermore, this result shows that hetero-functional graph theory does not suffer from any of the previously identified modeling constraints. Rather than continue discussion on this relatively small example, the next chapter (Chapter 5), demonstrates hetero-functional graph theory on a hypothetical smart city infrastructure system with features that more closely resemble the complexity of real-life city infrastructure.

Table 4.6 Summary of hetero-functional graph theory

Conceptual elements	Mathematical elements	Mathematical notation
(A) System Concept	System Form: Physical Resources	$R_P = B_S \cup H,\ B_S = M \cup H$
	System Function: Processes	$P = P_\mu \cup (P_\gamma \times P_\eta)$
	System Knowledge Base	$J_S = \left[\begin{array}{c\|c} J_M & \mathbf{0} \\ \hline \multicolumn{2}{c}{\leftarrow J_{\bar{H}} \rightarrow} \end{array}\right]$, where $J_{\bar{H}} = \left[J_\gamma \otimes \mathbb{1}^{\sigma(P_\eta)}\right] \cdot \left[\mathbb{1}^{\sigma(P_\gamma)} \otimes J_H\right]$
	System Constraints Matrix	$K_S = \left[\begin{array}{c\|c} K_M & \mathbf{0} \\ \hline \multicolumn{2}{c}{\leftarrow K_{\bar{H}} \rightarrow} \end{array}\right]$
	System Degrees of Freedom	$DOF_S = \sum_v^{\sigma(R)} \sum_w^{\sigma(P)} [J_S \ominus K_S](w, v) = \sum_v^{\sigma(R)} \sum_w^{\sigma(P)} A_S(w, v)$
(B) Hetero-functional Adjacency Matrix	System Sequence Knowledge Base	$J_\rho = A_S^V A_S^{VT} = \left[J_S \cdot \bar{K}_S\right]^V \left[J_S \cdot \bar{K}_S\right]^{VT}$
	System Sequence Constraints Matrix	K_ρ. Please refer to Table 4.3.
	Hetero-functional Adjacency Matrix	$A_\rho = J_\rho \ominus K_\rho$, or when projecting: $\tilde{A}_\rho = \mathbb{P}_S A_\rho \mathbb{P}_S^T$
	System Sequence Degrees of Freedom	$DOF_\rho = \sum_{\chi_1}^{\sigma(R)\sigma(P)} \sum_{\chi_2}^{\sigma(R)\sigma(P)} \left[A_\rho\right](\chi_1, \chi_2) = \sum_{\psi_1}^{\sigma(\mathcal{E}_S)} \sum_{\psi_2}^{\sigma(\mathcal{E}_S)} \left[\tilde{A}_\rho\right](\psi_1, \psi_2)$
(C) Controller Agency Matrix	Physical Resources	$R_P = B_S \cup H,\ B_S = M \cup H$
	(Cyber-) Resources	$R = R_P \cup Q$, where cyber-resources: $Q = Q_I \cup Q_D$
	Controller Agency Matrix	A_Q
(D) Controller Adjacency Matrix	Controller Adjacency Matrix	A_C
(E) Service as Operand Behavior	Services	$L = \{l_1, \ldots, l_{\sigma(L)}\}$
	Service Activities	$e_{xl_i} \in \mathcal{E}_{l_i}$
	Service String and Service Petri net	$z_{l_i} = e_{x_1 l_i} e_{x_2 l_i} \ldots e_{x_{\sigma(\mathcal{E}_{l_i})} l_i}$, and $N_{l_i} = \{S_{l_i}, \mathcal{E}_{l_i}, M_{l_i}, W_{l_i}, Q_{l_i}\}$

(continued)

Table 4.6 (continued)

Conceptual elements	Mathematical elements	Mathematical notation
(F) Service Feasibility Matrix	Service Transformation Feasibility Matrix	$\Lambda_{ML} = \bigvee_{i}^{\sigma(L)} \left[\left[\mathbb{1}^{\sigma(\mathcal{E}_{l_i})T} \odot \Lambda_{\mu i}\right]^T \mathbb{1}^{\sigma(M)T}\right]$
	Service Transportation Feasibility Matrix	$\Lambda_{HL} = \bigvee_{i}^{\sigma(L)} \left[\left[\mathbb{1}^{\sigma(\mathcal{E}_{l_i})T} \odot \Lambda_{\gamma i} \otimes \mathbb{1}^{\sigma(P_\eta)T}\right]^T \mathbb{1}^{\sigma(R)T}\right]$
	Service Line Feasibility Matrix	$\Lambda_{SL} = \left[\begin{array}{c\|c} \Lambda_{ML} & \mathbf{1} \\ \hline \multicolumn{2}{c}{\Lambda_{HL}} \end{array}\right] = \bigvee_{i}^{\sigma(L)} \Lambda_{Si}$
	Service Degrees of Freedom	$DOF_{LM} = \langle \Lambda_{ML} \cdot J_M, \bar{K}_M \rangle_F$ $DOF_{LH} = \langle \Lambda_{HL} \cdot J_{\bar{H}}, \bar{K}_{\bar{H}} \rangle_F$ $DOF_{LS} = \langle \Lambda_{SL} \cdot J_S, \bar{K}_S \rangle_F$

References

1. Borutzky, W. (2011). *Bond graph modelling of engineering systems*. New York: Springer.
2. Buede, D. M. (2009). *The engineering design of systems: Models and methods* (2nd ed.). Hoboken, NJ: Wiley.
3. Callaway, D. S., Newman, M. E. J., Strogatz, S. H., & Watts, D. J. (2000). Network robustness and fragility: Percolation on random graphs. *Physical Review Letters, 85*, 5468–5471. http://link.aps.org/doi/10.1103/PhysRevLett.85.5468
4. Crawley, E., Cameron, B., & Selva, D. (2015). *System architecture: Strategy and product development for complex systems*. Upper Saddle River, NJ: Prentice Hall Press.
5. David, R., & Alla, H. (2010). *Discrete, continuous, and hybrid Petri nets*. Berlin, Heidelberg: Springer.
6. De Weck, O. L., Roos, D., & Magee, C. L. (2011). *Engineering Systems: Meeting Human Needs in a Complex Technological World*. Cambridge, MA: MIT Press. http://www.knovel.com/knovel2/Toc.jsp?BookID=4611http://mitpress-ebooks.mit.edu/product/engineering-systems
7. Eppinger, S. D., & Browning, T. R. (2012). *Design structure matrix methods and applications*. Cambridge, MA: MIT Press.
8. Farid, A. M. (2007). *Reconfigurability Measurement in Automated Manufacturing Systems*. Ph.D. dissertation, University of Cambridge Engineering Department Institute for Manufacturing. http://engineering.dartmouth.edu/liines/resources/Theses/IEM-TP00.pdf
9. Farid, A. M. (2008). Facilitating ease of system reconfiguration through measures of manufacturing modularity. *Proceedings of the Institution of Mechanical Engineers, Part B (Journal of Engineering Manufacture) – Invited Paper, 222*(B10), 1275–1288. http://dx.doi.org/10.1243/09544054JEM1055
10. Farid, A. M. (2008). Product degrees of freedom as manufacturing system reconfiguration potential measures. *International Transactions on Systems Science and Applications – Invited Paper, 4*(3), 227–242. http://engineering.dartmouth.edu/liines/resources/Journals/IEM-J04.pdf
11. Farid, A. M. (2013). An axiomatic design approach to non-assembled production path enumeration in reconfigurable manufacturing systems. In *2013 IEEE International Conference on Systems Man and Cybernetics, Manchester, UK* (pp. 1–8). http://dx.doi.org/10.1109/SMC.2013.659
12. Farid, A. M. (2014). Static resilience of large flexible engineering systems: Part I – Axiomatic design model. In *4th International Engineering Systems Symposium* (pp. 1–8). Hoboken, NJ: Stevens Institute of Technology. http://engineering.dartmouth.edu/liines/resources/Conferences/IES-C37.pdf

13. Farid, A. M. (2014). Static resilience of large flexible engineering systems: Part II – Axiomatic design measures. In *4th International Engineering Systems Symposium* (pp. 1–8). Hoboken, NJ: Stevens Institute of Technology. http://engineering.dartmouth.edu/liines/resources/Conferences/IES-C38.pdf
14. Farid, A. M. (2015). Static resilience of large flexible engineering systems: axiomatic design model and measures. *IEEE Systems Journal, 99*, 1–12. http://dx.doi.org/10.1109/JSYST.2015.2428284
15. Farid, A. M. (2015). Multi-agent system design principles for resilient coordination and control of future power systems. *Intelligent Industrial Systems, 1*(3), 255–269. http://dx.doi.org/10.1007/s40903-015-0013-x
16. Farid, A. M. (2016). A hybrid dynamic system model for multi-modal transportation electrification. *IEEE Transactions on Control System Technology, 99*, 1–12. http://dx.doi.org/10.1109/TCST.2016.2579602
17. Farid, A. M. (2016). An engineering systems introduction to axiomatic design. In A. M. Farid & N. P. Suh (Eds.), *Axiomatic design in large systems: Complex products, buildings & manufacturing systems* (Chap. 1, pp. 1–47). Berlin, Heidelberg: Springer. http://dx.doi.org/10.1007/978-3-319-32388-6
18. Farid, A. M. (2017). Electrified transportation system performance: Conventional vs. online electric vehicles. In N. P. Suh & D. H. Cho (Eds.), *The on-line electric vehicle: Wireless electric ground transportation systems* (Chap. 20, pp. 279–313). Berlin, Heidelberg: Springer. http://engineering.dartmouth.edu/liines/resources/Books/TES-BC05.pdf
19. Farid, A. M. (2017). Measures of reconfigurability and its key characteristics in intelligent manufacturing systems. *Journal of Intelligent Manufacturing, 28*(2), 353–369. http://dx.doi.org/10.1007/s10845-014-0983-7
20. Farid, A. M., & McFarlane, D. C. (2006). A development of degrees of freedom for manufacturing systems. In *IMS'2006: 5th International Symposium on Intelligent Manufacturing Systems: Agents and Virtual Worlds, Sakarya, Turkey* (pp. 1–6). http://engineering.dartmouth.edu/liines/resources/Conferences/IEM-C02.pdf
21. Farid, A. M., & McFarlane, D. C. (2007). A design structure matrix based method for reconfigurability measurement of distributed manufacturing systems. *International Journal of Intelligent Control and Systems Special Issue – Invited Paper, 12*(2), 118–129. http://engineering.dartmouth.edu/liines/resources/Journals/IEM-J02.pdf
22. Farid, A. M., & McFarlane, D. C. (2008). Production degrees of freedom as manufacturing system reconfiguration potential measures. *Proceedings of the Institution of Mechanical Engineers, Part B (Journal of Engineering Manufacture) – Invited Paper, 222*(B10), 1301–1314. http://dx.doi.org/10.1243/09544054JEM1056
23. Farid, A. M., & Ribeiro, L. (2015). An axiomatic design of a multi-agent reconfigurable mechatronic system architecture. *IEEE Transactions on Industrial Informatics, 11*(5), 1142–1155. http://dx.doi.org/10.1109/TII.2015.2470528
24. Fischer, M. J., & Meyer, A. R. (1971). Boolean matrix multiplication and transitive closure. In *12th Annual Symposium on Switching and Automata Theory, 1971* (pp. 129–131). New York, NY: IEEE.
25. Friedenthal, S., Moore, A., & Steiner, R. (2011). *A practical guide to SysML: The systems modeling language* (2nd ed.) Burlington, MA: Morgan Kaufmann.
26. Holme, P., & Saramäki, J. (2012). Temporal networks. *Physics Reports, 519*(3), 97-125.
27. Hoyle, D. (1998). *ISO 9000 pocket guide*. Oxford, Boston: Butterworth-Heinemann [Online]. Available: http://www.loc.gov/catdir/toc/els033/99163006.html
28. Khayal, I. S., & Farid, A. M. (2015). Axiomatic design based volatility assessment of the Abu Dhabi Healthcare Labor Market. *Journal of Enterprise Transformation, 5*(3), 162–191. http://dx.doi.org/10.1080/19488289.2015.1056449
29. Khayal, I. S., & Farid, A. M. (2016). The need for systems tools in the practice of clinical medicine. *Systems Engineering, 20*(1), 3–20. http://dx.doi.org/10.1002/sys.21374
30. Khayal, I. S., & Farid, A. M. (2017). Architecting a system model for personalized healthcare delivery and managed individual health outcomes. *Complexity, 1*(1), 1–25. https://doi.org/10.1155/2018/8457231

31. Khayal, I., & Farid, A. M. (2017). A dynamic model for a cyber-physical healthcare delivery system with human agents. In *2017 IEEE International Conference on Systems, Man, and Cybernetics (SMC2017), Intelligent Industrial System Special Session, Banff, Canada, October 2017*.
32. Khayal, I., & Farid, A. M. (2018). Designing patient-oriented healthcare services as systems of systems. In *2018 IEEE Systems of Systems Conference, Paris, France*.
33. Lloyd, S. (2000). Ultimate physical limits to computation. *Nature, 406*(6799), 1047.
34. Newman, M. E. (2003). The structure and function of complex networks. *SIAM Review, 45*(2), 167–256.
35. Pavella, M., Ernst, D., & Ruiz-Vega, D. (2000). *Transient Stability of Power Systems A Unified Approach to Assessment and Control*. Boston: Kluwer Academic Publishers.
36. Pine, J. (1993). Mass customizing products and services. *Planning Review, 21*, 6–55.
37. Pine, J. B. (1993). *Mass customization: The New Frontier in business competition*. Cambridge, MA: Harvard Business School Press.
38. Popova-Zeugmann, L. (2013). *Time Petri nets*. Berlin, Heidelberg: Springer.
39. Schoonenberg, W. C., & Farid, A. M. (2015). A dynamic production model for industrial systems energy management. In *2015 IEEE International Conference on Systems Man and Cybernetics, Hong Kong* (pp. 1–7). http://dx.doi.org/10.1109/SMC.2015.14
40. Schoonenberg, W. C., & Farid, A. M. (2017). A dynamic model for the energy management of microgrid-enabled production systems. *Journal of Cleaner Producation, 1*(1), 1–10.
41. SE Handbook Working Group (2015). *Systems engineering handbook: A guide for system life cycle processes and activities*. San Diego, CA: International Council on Systems Engineering (INCOSE).
42. Shabana, A. A. (1998). *Dynamics of multibody systems* (2nd ed.) Cambridge: Cambridge University Press.
43. Silveira, G. D., Borenstein, D., Fogliatto, H. S., Da Silveira, G., & Fogliatto, F. S. (2001). Mass customization: Literature review and research directions. *International Journal of Production Economics, 72*(1), 1–13.
44. Smith, S., Jiao, R., & Chu, C.-H. (2012). Editorial: Advances in mass customization. *Journal of Intelligent Manufacturing, 24*(5), 873–876.
45. Smith, S., Smith, G. C., Jiao, R., & Chu, C.-H. (2012). Mass customization in the product life cycle. *Journal of Intelligent Manufacturing, 24*(5), 877–885.
46. Suh, N. P. (2001). *Axiomatic design: Advances and applications*. Oxford: Oxford University Press.
47. van der Wardt, T. J., & Farid, A. M. (2017). A hybrid dynamic system assessment methodology for multi-modal transportation-electrification. *Energies, 10*(5), 653. http://dx.doi.org/10.3390/en10050653
48. Viswanath, A., Baca, E. E. S., & Farid, A. M. (2014). An axiomatic design approach to passenger itinerary enumeration in reconfigurable transportation systems. *IEEE Transactions on Intelligent Transportation Systems, 15*(3), 915–924. http://dx.doi.org/10.1109/TITS.2013.2293340
49. Warner, S. (1990). *Modern algebra*. Chelmsford, MA: Courier Corporation.
50. Weilkiens, T. (2007). *Systems engineering with SysML/UML modeling, analysis, design*. Burlington, MA: Morgan Kaufmann.

Chapter 5
Modeling Interdependent Smart City Infrastructure Systems with HFGT

This book develops a Hetero-functional Graph Theory for structural modeling of smart city infrastructure. The previous chapter provided an exposition of hetero-functional graph theory in terms of its constituent mathematical models. This chapter now applies the theory to an interdependent smart city infrastructure test case called *"Trimetrica."* First, the role of test cases in smart city development is discussed in Section 5.1. Thereafter, the Trimetrica test case is introduced in Section 5.2. Then Sections 5.3 through 5.9 subsequently discuss the seven elements of hetero-functional graph theory.

5.1 The Role of Test Cases in Smart City Development

Test cases have served an invaluable role in the development of infrastructure systems. While data collected from real-life cities is invaluable for making conclusions about certain smart city instances, it is often unclear how modeling, analysis, and simulation methods used in one case study can be applied to other cities. Any one smart city may not be representative of the class of smart cities. Indeed, the process of data collection itself comes with inherent ontological assumptions that are often overlooked. Reconsider the discussion from Chapter 3. Figure 3.3 (on page 25) shows that relying on data-driven from a single smart city instance may violate the ontological properties of completeness and lucidity. Stated differently, the smart city data used in the model may not address all smart city abstractions (i.e., violation of completeness). For example, the footnote from Example 3.2 (on page 31) mentions that power flow analysis data fails to capture the data from the lead lines between generators and substations. The smart city data used in the model may also be applied to multiple smart city abstractions (i.e., violation of lucidity). For example, the footnote from Example 3.2 (on page 31) mentions that power flow analysis data uses the same data to refer to both power generation and

W. C. H. Schoonenberg et al., *A Hetero-functional Graph Theory for Modeling Interdependent Smart City Infrastructure*,
https://doi.org/10.1007/978-3-319-99301-0_5

energy storage facilities. Test cases resolve many[1] of these concerns because their meta-data implies an underlying reference architecture which may be ontologically critiqued for generality.

Many infrastructure engineering fields have recognized these concerns from a practical perspective. The power systems engineering field developed strong rationales for the usage of test cases as early as the 1970s [17] and has since developed several test case repositories [1, 12]. Water distribution engineers have used the famous "Anytown" network [3]. In both cases, these test cases allowed the development of rigorous numerical methods without compromising the physical security of sharing data about critical infrastructure [4, 7, 13, 18]. Despite a general trend towards transportation system instances, some transportation system engineers have advocated the use of test cases to understand the fundamental properties of transportation dynamics [19]. Many of these developments arguments were presented in the development of "Symmetrica" as a test case for the development transportation-electrification research [9]. It is on this foundation that "Trimetrica" is developed as an interdependent smart city infrastructure.

5.2 Smart City Test Case: Trimetrica

Trimetrica contains three infrastructure systems: a power system, a transportation system, and a water distribution system. The first two are drawn from the "Symmetrica" test case [9]. The third is derived from the "Anytown" water network [3]. These systems are chosen as three critical infrastructures within any smart city. They also include both physical and cyber-resources. Additionally, their dynamic behaviors are well-known. Therefore, this test case may, in the future, provide a basis for dynamic cyber-physical modeling of smart cities. The topological lay-out is presented in Figure 5.1 and the associated data is publicly available [10]. This section first discusses each of the individual infrastructure systems. Secondly, it details the connection between the infrastructure systems, and thirdly the section highlights the "architectural" assumptions in this test case.

The *electric power system* is based on the IEEE 201-bus test case [8, 15] and is depicted in Figure 5.1b. The power system consists of a single power generation bus, 200 load buses, and 200 branches. The power system supplies electric power to the consumption nodes, which in many cases are connected to the other infrastructure systems. If a power node is connected to another system, it is assumed that it maintains a separate power consumption function. The power system provides the service "delivery of electric power at 132 kV," which upon arrival may be used

[1] While reference architectures overcome doubts about how representative a certain instantiated architecture may be, their meta-data may still violate the four ontological properties. The examples provided above indicate the degeneracy of power flow analysis data despite its widespread use.

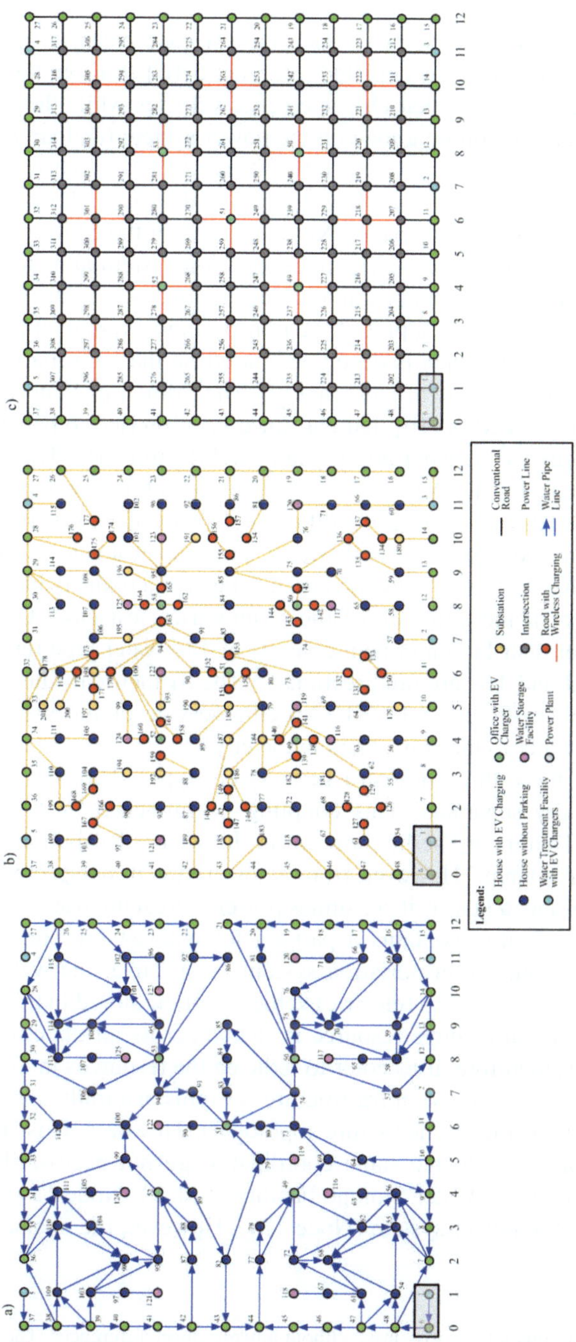

Fig. 5.1 Topological depiction of the Trimetrica smart city infrastructure test case: the networks are topologically superimposed. **(a)** Topology of (electrified) water distribution system. **(b)** Topology of electric power system. **(c)** Topology of electrified transportation system

directly or to support the transformation and storage of other processes at the consumption node.

The *transportation system* in the Trimetrica test case consists of 169 nodes, structured in a grid-like shape, as depicted in Figure 5.1c [9]. From the 169 nodes, 53 nodes are charging stations that charge electric vehicles while they are parked. These charging locations impose a load on the power system (and are thus a cross-layer connection). The nodes are connected by 624 roads, 104 of which are charging roads that charge electric vehicles as they drive over the road. The charging roads also impose a load on the power system. Note that each of the edges in Figure 5.1c represents two unidirectional roads. The transportation system aims to provide the service "transport person from an origin to a destination," where the person uses a vehicle, and the vehicle remains a part of the transportation system at all times.

The *water distribution system* is developed based on the University of Exeter's Anytown water distribution network test case [3]. The Trimetrica Water Distribution System was first introduced in [16], but has since been revised to better match the electric power system and transportation system architecture. The Trimetrica water distribution system scales the Anytown network by a factor 5, arriving at 5 water treatment facilities, 10 water storage facilities, and 110 water consumption nodes. These consumption nodes provide both warm and cold water to the consumers. The structure of the Anytown network has not been maintained identically, but has been revised to fit the structure of the Symmetrica test case. Consequently, the Trimetrica water distribution system consists of 186 water pipe lines, of which 19 are bidirectional and 167 are unidirectional.[2] The Trimetrica water distribution system is presented in Figure 5.1a. The specific service of the water system is delivering potable water to the consumers in the city.

In a smart city, these three infrastructure system types share resources, rather than being physically separated. This coupling is mimicked in Trimetrica. For example, a Water Treatment Facility appears in the water distribution system as a "water generating resource." However, from a power system perspective, the water treatment facility is a load bus, as it consumes power to treat the water. Additionally, the water treatment facility contains a parking lot with an EV charging facility for its employees. From a transportation system perspective, the water treatment facility is a buffering resource, as (electric) vehicles can be stored in the parking lot. This example shows that a single resource is shared across all three infrastructure systems. Figure 5.1, therefore, uses color to indicate the resources in Trimetrica.

The resources shared across infrastructure systems naturally have the same coordinates in each of the infrastructure topologies. However, it is important to note that two resources with the same coordinates are not necessarily the same resource. For example, the Water Storage Facility at (4, 9) (Node 124) is the same resource in the electric power system (Node 124). However, the intersection in the

[2]The water pipes are assumed to be loss-less, without any elevation differences. The water system pressure is assumed to be maintained by the water treatment facilities and the water storage facilities, rather than the use of pumps in the pipe lines.

transportation system, Node 288 at (4, 9), is separate from the water storage facility and consequently not the same resource.

Some resources, however, have a different representation in the infrastructure topologies. The transportation system uses edges to represent roads between intersections. Each of these edges represents two separate roads in opposite directions. However, when these roads facilitate wireless charging of the vehicles, they share a single load bus in the electric power system. When the systems are discussed independently, they contain three separate resources. However, as a result of explicitly sharing resources across infrastructure systems, the three resources aggregate into a single resource. This resource is presented as an edge in the transportation system, and as a node in the electric power system. An example is Node 126 in the electric power system, which is also the edge from Node 203 to Node 214, which in turn represents both unidirectional roads between Nodes 203 and 214. Table 5.1 provides an overview of all system resources that appear

Table 5.1 Resources in Trimetrica with associated infrastructure system and controller type

		Infrastructure Type			Controller Type			
#	Resources	Transportation System	Electric Power System	Water Distribution System	End User	Water Utility	Power Utility	Count
1	Water treatment facility w/ EV charger	■	■	■		■		5
2	House w/ EV charger	■	■	■	■			43
3	Office w/ EV charger	■	■	■	■			5
4	House w/o parking		■	■	■			62
5	Water storage facility		■	■		■		10
6	Electrified road	■	■		■			52
7	Power plant		■				■	1
8	Substation		■				■	23
9	Intersection	■			■			116
10	Conventional road	■			■			520
11	Electric power line		■				■	200
12	Directed water pipeline			■		■		167
13	Undirected water pipeline			■		■		19

in Trimetrica. The table also identifies the infrastructure system(s) in which the resources appear.

For reasons of clarity, the cyber-resources are not shown in Figure 5.1. Table 5.1 also provides an overview of the cyber-resources, and their relationship to Trimetrica's resources. Note that the Electric Power System and the Water Distribution System generation and transportation resources are controlled centrally. In contrast, the transportation system is controlled solely by the end users.

The Trimetrica smart city infrastructure test case is designed to provide a platform to compare modeling methods of smart city infrastructure systems. As discussed before in Chapter 3, these infrastructure systems are Large Flexible Engineering Systems both separately and when combined. Hetero-functional Graph Theory can thus be used to model the Trimetrica test case, as demonstrated in the remainder of this chapter.

Given its size, visualizing the Trimetrica test case is often challenging. Therefore, the book uses both full and partial visualizations to support the mathematics. The topological lay-out of the full model is depicted in Figure 5.1. The partial model is independently depicted in Figure 5.2.

After this introduction of the Trimetrica test case, the chapter now continues by modeling Trimetrica using Hetero-functional Graph Theory. The chapter discusses all seven elements of Hetero-functional Graph Theory sequentially to derive a Hetero-functional System Adjacency Matrix as a complete, mathematical model of Trimetrica.

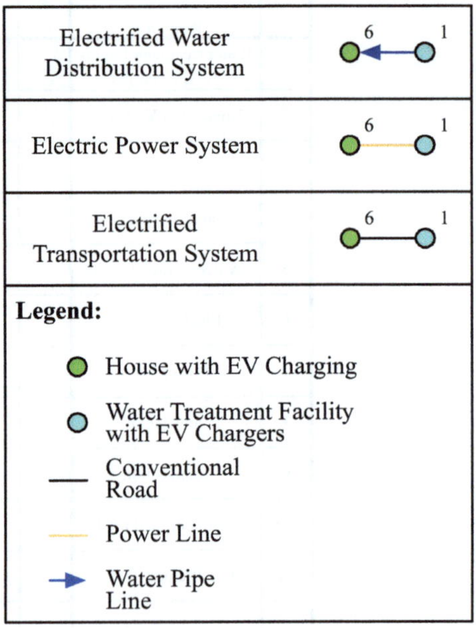

Fig. 5.2 Partial topological depiction of the Trimetrica smart city infrastructure test case: the networks are topologically superimposed

5.3 System Concept

System concept is introduced in Section 4.1 as the mapping of system function onto system form. This section discusses system concept for the Trimetrica test case. Trimetrica consists of three interfacing infrastructure systems, as introduced in Section 5.2. It is now critical to define system concept accurately across the disciplinary systems to derive the correct System Adjacency Matrix in the end. The section first discusses the resources, followed by the processes. Finally, this section maps the processes to the resources to calculate the smart city knowledge base.

5.3.1 *Smart City Resources*

The three smart city infrastructures both independently and collectively adhere to the definition of a Large Flexible Engineering System (LFES). Therefore, to derive the smart city resources, this section relies on the resource architecture of Large Flexible Engineering Systems as discussed in Section 4.1, in Figure 4.4 on Page 47.

In the Trimetrica Case Study, the LFES resource structure is specialized into three types: a transportation, an electric power, and a water distribution resource structure. By virtue of this specialization, the resource structure of the three infrastructure systems equals the structure of the LFES. This is best explained using Figure 5.3, in which the SysML representation of the LFES resource meta-architecture is depicted aside the specialized resource architectures of the transportation system, electric power system, and the water distribution system. Each LFES consists of a set of resources R. Similar to the specialization of the LFES meta-architecture, the set of resources R is specialized into transportation system resources R_T, electric power system resources R_E, and water distribution system resources R_W. Mathematically, $R = R_T \cup R_E \cup R_W$.

Furthermore, Figure 5.3 shows the atomic class structure of the set of LFES resources: $R = M \cup B \cup H \cup Q$. Similarly, the atomic class structure of the infrastructure systems is:

- Transportation System: $R_T = M_T \cup B_T \cup H_T \cup Q_T$
- Electric Power System: $R_E = M_E \cup B_E \cup H_E \cup Q_E$
- Water Distribution System: $R_W = M_W \cup B_W \cup H_W \cup Q_W$

Note that these resource classes are also specializations of the LFES resources:

- Transformation Resources $M = M_T \cup M_E \cup M_W$
- Independent Buffers $B = B_T \cup B_E \cup B_W$
- Transportation Resources $H = H_T \cup H_E \cup H_W$
- Cyber Resources $Q = Q_T \cup Q_E \cup Q_W$

The set of atomic resource classes in Figure 5.3 specializes each of the infrastructure systems individually. This specialization does not specify if resources interface

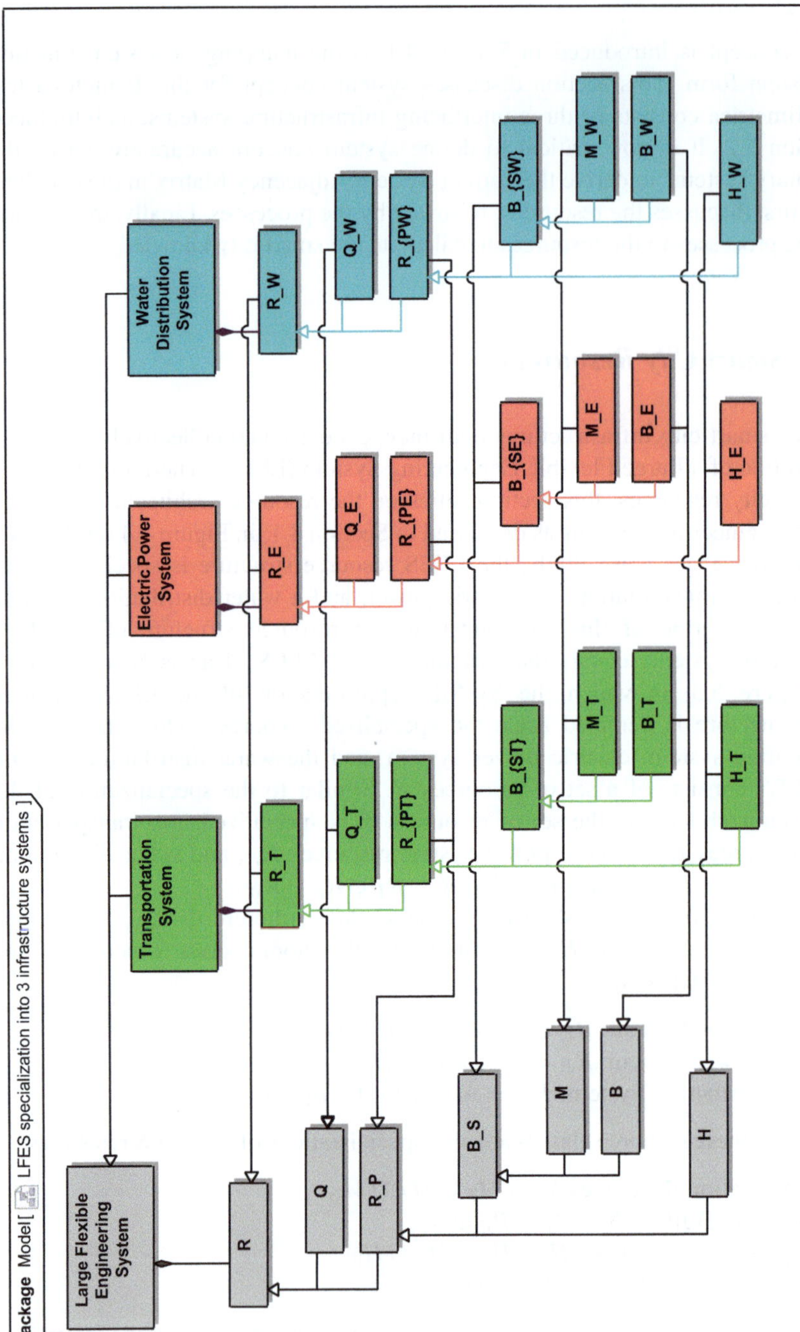

Fig. 5.3 SysML specialization of three infrastructure systems relative to the LFES meta-architecture

or interact across infrastructure systems. Therefore, *interface classes* are introduced as a specialization that distinguishes resource classes based on their interfaces or interactions with other infrastructure types. Figure 5.4 shows such a specialization. The atomic resource classes are specialized into four groups of interface classes.

- The first group is *conventional*, indicated by subscript c. The conventional resources do not interact across disciplinary boundaries. For example, H_{TC} is a conventional transportation resource in the transportation system.
- The second and third groups are resources that interact with one *other* infrastructure system. For example, M_{WE} is a water distribution system transformation resource that interfaces with the electric power system.
- The fourth group contains resources that interact with two *other* infrastructure systems. For example, M_{ETW} is an electric power system resource that interfaces with both the transportation and the water distribution system.

Note that each of the subscripts indicates the infrastructure interface type. The first subscript indicates its original infrastructure type, and the second and third subscript are arbitrarily sequenced.

The interface classes in Figure 5.4 are still defined within a specified infrastructure system. However, some resources are part of multiple infrastructure systems. Therefore, the interface classes must be specialized into multi-operand resource classes. These multi-operand resource classes inherit their properties from multiple interface classes in distinct infrastructure systems. The multi-operand resource classes are thus truly part of multiple infrastructure systems. Figure 5.5 presents an overview of the multi-operand resource classes that appear in Trimetrica.[3]

It is now important to note that if hetero-functional graph theory is applied to another test case or system, these specializations will be different. For example, if Trimetrica did not have three infrastructure systems but four infrastructure systems, the number of interface classes would be much larger. Each of the atomic resource classes is currently specialized into four possible combinations.[4] With four infrastructure systems the number of possible combinations becomes eight. Similarly, the number of potential multi-operand resource classes as combinations of interface classes will increase rapidly by virtue of an added layer.

However, Trimetrica uses only a subset of the available interface classes and multi-operand resource classes. Trimetrica contains 13 resource types, as specified in Table 5.1. These resource types are now matched to the interface and multi-operand resource classes:

[3]The name of multi-operand resource classes uses the following convention: its name is a combination of each of its parent interface classes. The sequence of the combination is: first the transformation resource, then the buffer, and last the transportation resource. For example, $M_E\&B_W$ is a multi-operand resource that inherits from the interface class M_{EW} in the electric power system and B_{WE} in the water distribution system.

[4]Figure 5.4 presents 36 possible resource types. In all three systems, the transformation resources, independent buffers, and transportation resources are specialized as four interface classes.

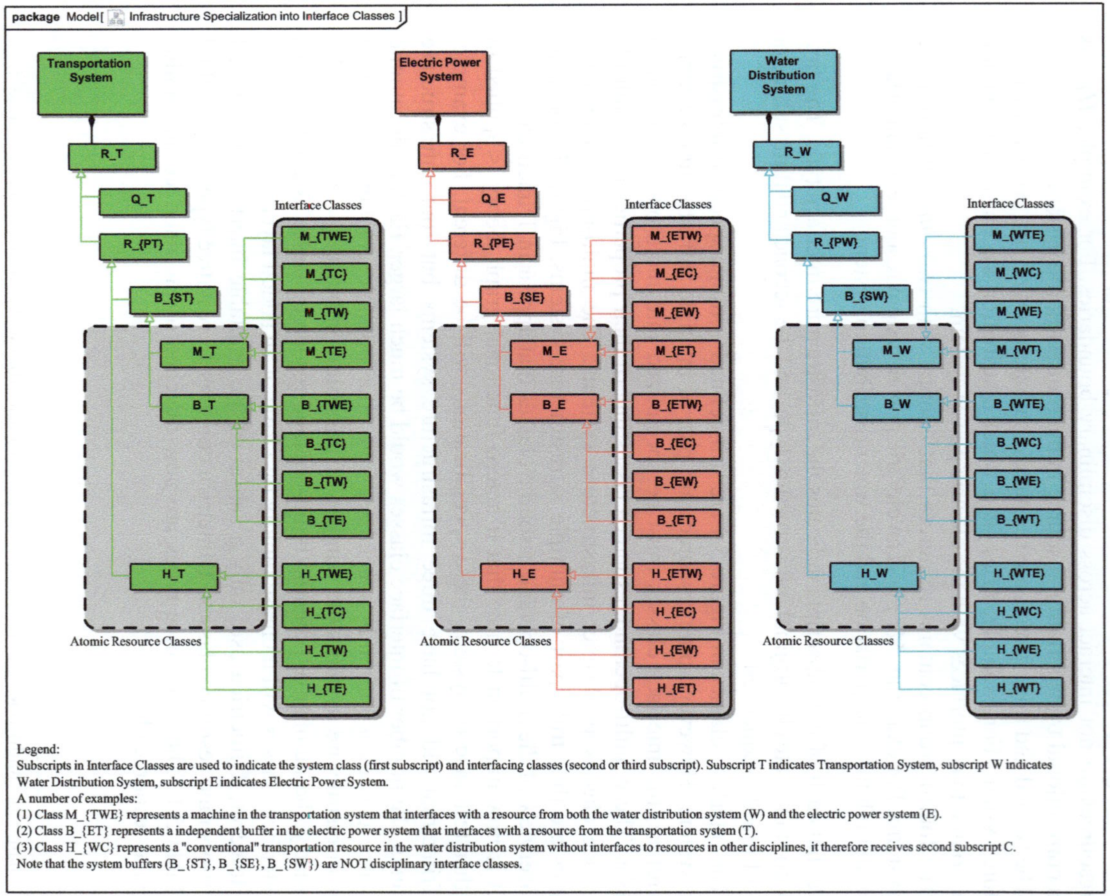

Fig. 5.4 Full SysML overview of the disciplinary system class structure with specialization of atomic disciplinary classes into the complete set of interface classes

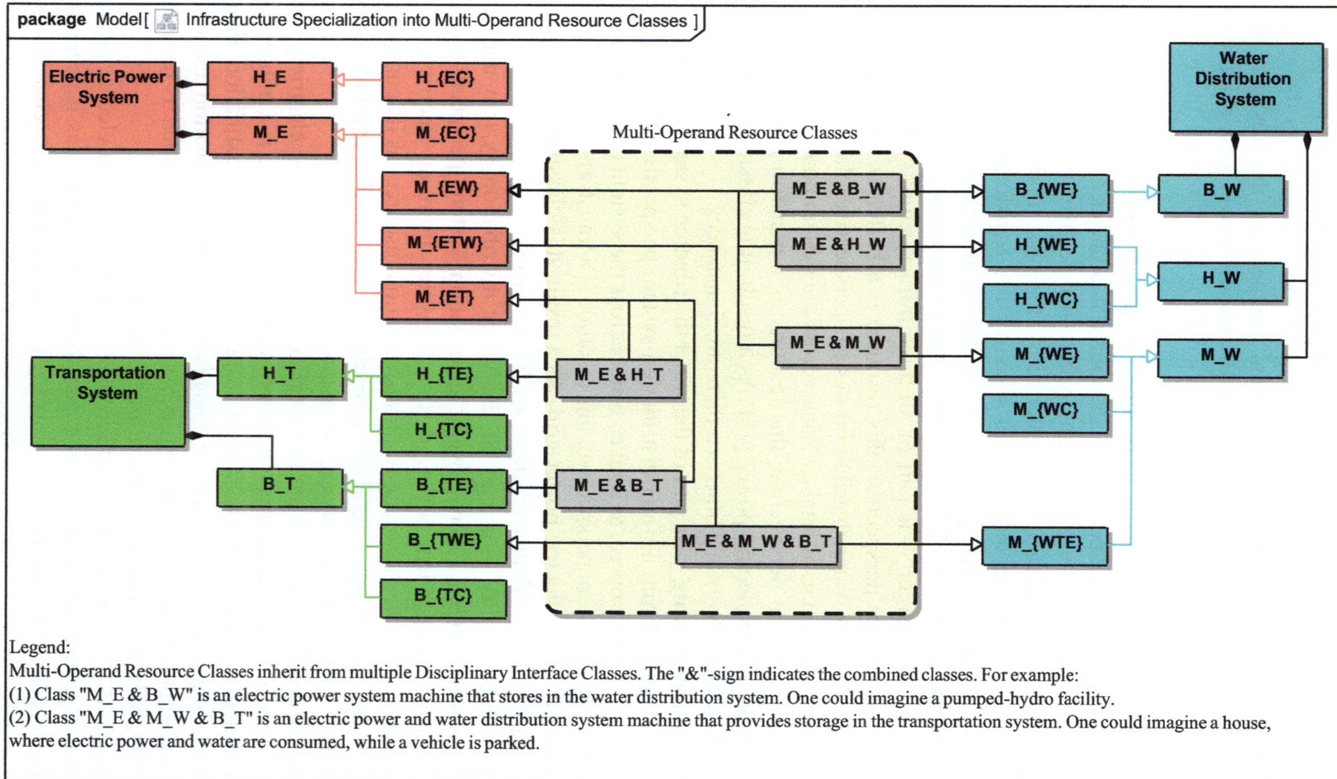

Fig. 5.5 SysML block definition diagram of the Trimetrica infrastructure systems specialized to define the multi-operand resources that allow the disciplinary systems to interface

1. "Water treatment facility with EV charger"—is part of all three infrastructure systems. From a transportation system perspective, this resource is a buffer as it stores an EV temporarily. From a water distribution system perspective, this resource is a transformation resource, as it adds water across the system boundary. From an electric power system perspective, this resource is a transformation resource, as it is a load on the electric power grid. As a result, it is classified as resource class: $(M_E \& M_W \& B_T)$. In Trimetrica there are 5 instances.
2. "House with EV charger"—is part of all three infrastructure systems. From a transportation system perspective, this resource is a buffer as it stores an EV temporarily. From a water distribution system perspective, this resource is a transformation resource, as it takes water across the system boundary. From an electric power system perspective, this resource is a transformation resource, as it is a load on the electric power grid. As a result, it is classified as resource class: $(M_E \& M_W \& B_T)$. In Trimetrica there are 43 instances.
3. "Office with EV charger"—is part of all three infrastructure systems. From a transportation system perspective, this resource is a buffer as it stores an EV temporarily. From a water distribution system perspective, this resource is a transformation resource, as it takes water across the system boundary. From an electric power system perspective, this resource is a transformation resource, as it is a load on the electric power grid. As a result, it is classified as resource class: $(M_E \& M_W \& B_T)$. In Trimetrica there are 5 instances.
4. "House without parking"—is part of the water distribution system and the electric power system. From a water distribution system perspective, this resource is a transformation resource, as it takes water across the system boundary. From an electric power system perspective, this resource is a transformation resource, as it is a load on the electric power grid. As a result, it is classified as resource class: $(M_E \& M_W)$. In Trimetrica there are 62 instances.
5. "Water storage facility"—is part of the water distribution system and the electric power system. From a water distribution system perspective, this resource is an independent buffer, as it stores water. From an electric power system perspective, this resource is a transformation resource, as its pumps are a load on the electric power grid. As a result, it is classified as resource class: $(M_E \& B_W)$. In Trimetrica there are 10 instances.
6. "Electrified road"—is part of the transportation system and the electric power system. From a transportation system perspective, this resource is a transportation resource as it transports the vehicle from its origin to its destination. From an electric power system perspective, this resource is a transformation resource, as it is a load on the electric power grid. As a result, it is classified as resource class: $(M_E \& H_T)$. In Trimetrica there are 52 instances.
7. "Power plant"—is part of the electric power system. From an electric power system perspective, this resource is a transformation resource, as it generates power and adds it to the electric power grid. As a result, it is classified as resource class: M_{EC}. In Trimetrica there is one instance.

8. "Substation"—is part of the electric power system. From an electric power system perspective, this resource is a transformation resource, as all substations consume at least a minor amount of power. Additionally, substations are conventionally modeled as load buses in power systems engineering. As a result, it is classified as resource class: M_{EC}. In Trimetrica there are 23 instances.
9. "Intersection"—is part of the transportation system. From a transportation system perspective, this resource is a buffer as it stores a vehicle temporarily while it redirects from one road to another. As a result, it is classified as resource class: B_{TC}. In Trimetrica there are 116 instances.
10. "Conventional road"—is part of the transportation system. From a transportation system perspective, this resource is a transportation resource as it transports the vehicle from its origin to its destination. As a result, it is classified as resource class: H_{TC}. In Trimetrica there are 520 instances.
11. "Electric power line"—is part of the electric power system. From an electric power system perspective, this resource is a transportation resource, as it transports electric power from its origin to its destination. As a result, it is classified as resource class: H_{EC}. In Trimetrica there are 200 instances.
12. "Directed water pipeline"—is part of the water distribution system. From a water distribution system perspective, this resource is a transportation resource, as it transports water from its origin to its destination. As a result, it is classified as resource class: H_{WC}. In Trimetrica there are 167 instances.
13. "Undirected water pipeline"—is part of the water distribution system. From a water distribution system perspective, this resource is a transportation resource, as it transports water from its origin to its destination. As a result, it is classified as resource class: H_{WC}. In Trimetrica there are 19 instances.

Figure 5.6 provides an overview of the resource structure in Trimetrica.[5] Additionally, the figure shows that the resource structure of the integrated smart city infrastructure system is a type of LFES. The set of Smart City Resources is: $R_{SC} = (M_E \& M_W \& B_T) \cup (M_E \& M_W) \cup (M_E \& B_W) \cup (M_E \& H_T) \cup M_{EC} \cup B_{TC} \cup H_{TC} \cup H_{EC} \cup H_{WC}$. These sets are mutually exclusive and collectively exhaustive as they contain all resources, without overlap. The 9 resource sets have the following sizes:

- $\sigma(M_E \& M_W \& B_T) = 53$. The set contains 5 water treatment facilities, 5 offices, and 43 houses, all with EV chargers.
- $\sigma(M_E \& M_W) = 62$. The set contains 62 houses without parking or EV chargers.
- $\sigma(M_E \& B_W) = 10$. The set contains 10 water storage facilities, without parking or EV chargers.
- $\sigma(M_E \& H_T) = 52$. The set contains 52 electrified roads, able to transport vehicles, and charge vehicles in both directions.[6]

[5]Note that the cyber-resources Q_{SC} are discussed in Section 5.5 on Page 131.

[6]Note that contrary to set H_{TC}, this set has one resource per edge with a bidirectional capability.

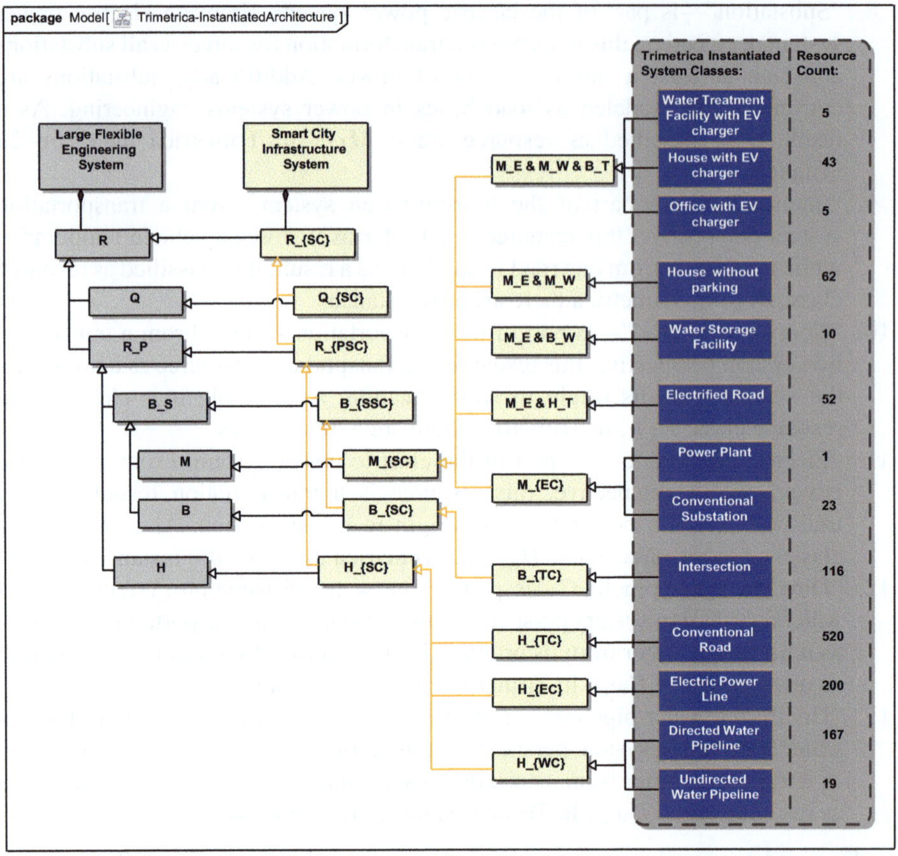

Fig. 5.6 SysML block definition diagram of the Trimetrica smart city infrastructure system as a specialization of the LFES meta-architecture: This figure shows that Trimetrica's smart city infrastructure system is a single system, rather than three separate systems. The three systems, each classified as an LFES in Figure 5.3, are reconciled into a single smart city infrastructure, of type LFES

- $\sigma(M_{EC}) = 24$. The set contains 23 conventional substations, that all impose an electric load on the power grid, and a power plant that generates electric power.
- $\sigma(B_{TC}) = 116$. The set contains 116 intersections.
- $\sigma(H_{TC}) = 520$. The set contains 520 conventional roads.
- $\sigma(H_{EC}) = 200$. The set contains 200 power lines.
- $\sigma(H_{WC}) = 186$. The set contains 167 directed water pipelines, and 19 undirected water pipelines.

The set of smart city resources R_{SC} has a size of 1223. The set of smart city buffers B_{SSC} has a size of 317, the set of smart city transformation resources M_{SC} has a size of 201, and the set of smart city independent buffers B_{SC} has a size of 116.

Lastly, the set of smart city transportation resources H_{SC} has a size of 906. The set of smart city system resources has been derived, and the next step is the derivation of the smart city system processes.

5.3.2 *Smart City Processes*

The definition of Trimetrica's System Processes follows an approach similar to the definition of the System Resources. First, the Large Flexible Engineering Systems' Meta-Architecture is introduced, to then discuss its specializations in the three infrastructure systems. Thereafter, the system processes are reconciled to define the final set of Trimetrica's system processes.

The LFES meta-architecture for system processes is presented using a SysML Activity Diagram. Figure 5.7 restates the meta-architecture as introduced in Section 4.1.1 on Page 40. The functional meta-architecture is a set of processes, classified as transformation processes, transportation processes, and holding processes. Their sequence is unrestricted as a result of a lack of general, cross-disciplinary constraints. For the specialized reference architectures, however, process sequence is constrained by the limitations specific to the reference architecture. This section introduces the design pattern for each of Trimetrica's three infrastructure disciplines separately. Thereafter, the design patterns are reconciled to create a triple-infrastructure design pattern. Trimetrica is an instantiation of this triple-infrastructure design pattern. The set of Trimetrica's instantiated system processes is therefore derived from its design pattern.

The *water distribution system* reference architecture is presented as an activity diagram in Figure 5.8. The operand crosses the system boundary when entering and leaving the system which requires transformation processes. A distinction is made

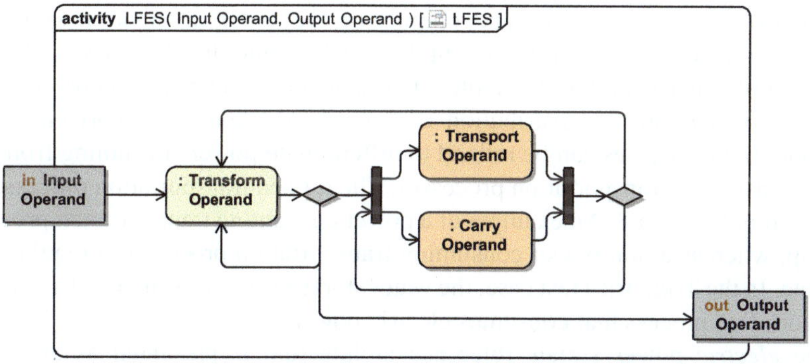

Fig. 5.7 Activity diagram of the LFES meta-architecture: the diamonds represent exclusive decisions. For example, after "Transform Operand" one of three options must be chosen: (1) "Transform Operand," (2) "Transport Operand with Carry Operand," or (3) End the sequence by creating: "Output Operand"

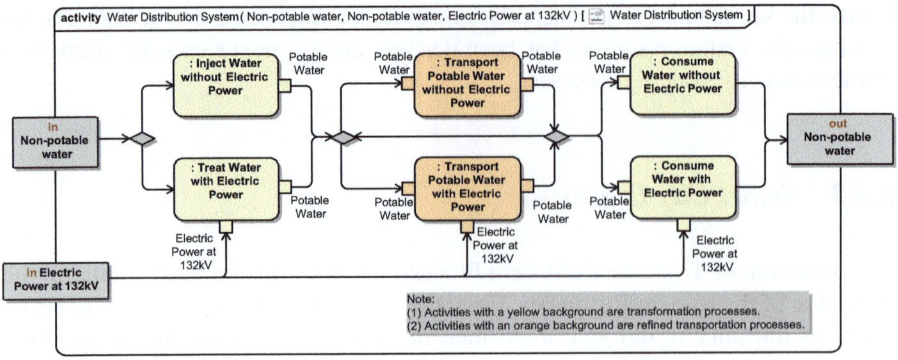

Fig. 5.8 Activity diagram of the water distribution system reference architecture

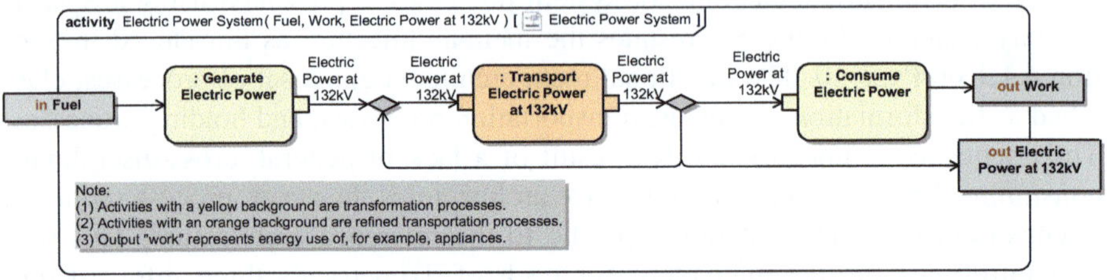

Fig. 5.9 Activity diagram of the electric power system reference architecture

between input and output transformations. The input and output transformations are classified as power consuming or non-power consuming. Examples of the former are (1) the extraction of ground water, and (2) hot water consumption. An example of the latter is a river that flows through a city. Note that the power consuming processes require the input of electric power which is transformed to perform work or generate heat.

The location of the inputs and outputs of the water distribution system are physically separated. Water treatment is usually centralized, whereas the consumption of water is highly distributed. Consequently, transportation processes are required to facilitate the distribution of water throughout the smart city. These transportation processes can be refined to differentiate power consuming from non-power consuming transportation processes. The refined transportation processes are presented in Figure 5.8. An example of a power consuming transportation process is a pump, whereas a non-power consuming transportation process is a regular water pipeline. In the Trimetrica test case, the water storage process is modeled as a refined transportation process that consumes electric power.

The *electric power system* reference architecture is presented as an activity diagram in Figure 5.9. Similar to the water distribution system, the electric power system operands enter and leave the system by virtue of transformation processes. These transformation processes are distinctive in type and location, and are therefore separated in the reference architecture as "generate electric power" (electric power entering the system) and "consume electric power" (electric power leaving the system). Note that the activity diagram indicates that some power is crossing the system boundary as an output to perform work in the other infrastructure systems. This emphasizes the need for a holistic assessment of the water distribution system, electric power system, and the transportation system. Transportation processes necessarily occur between generation and consumption, as these processes are physically separated. Electric power is conventionally expressed as a "per unit" voltage relative to the designed line capacity. Consequently, there is no need to differentiate between the holding processes in the electric power system in this example.

The *electrified transportation system* design pattern is presented as an activity diagram in Figure 5.10. The transportation system's core function is to transport vehicles to locations in the smart city. However, the transportation system may also receive vehicles from other locations, or accommodate vehicles that leave the city. The input and output processes are accommodated by transformation processes. The transportation processes are refined to represent the types of transportation with electric vehicles as the operand. Electric vehicles discharge while driving, but the impact of parking on the battery capacity is negligible. Additionally, the electric vehicles need to be charged by wire when they are stationary, or wirelessly if they are charged while driving. The power consumption of these charging processes is facilitated by the electric power system, and requires the input of electric power. Similar to the water distribution system, the charging processes transform the electric power into (portable) charge.

The *triple-infrastructure smart city* design pattern is a combination of each of the infrastructure's individual design patterns. The design pattern is displayed as an activity diagram in Figure 5.11. Note that electric power is consumed internally,

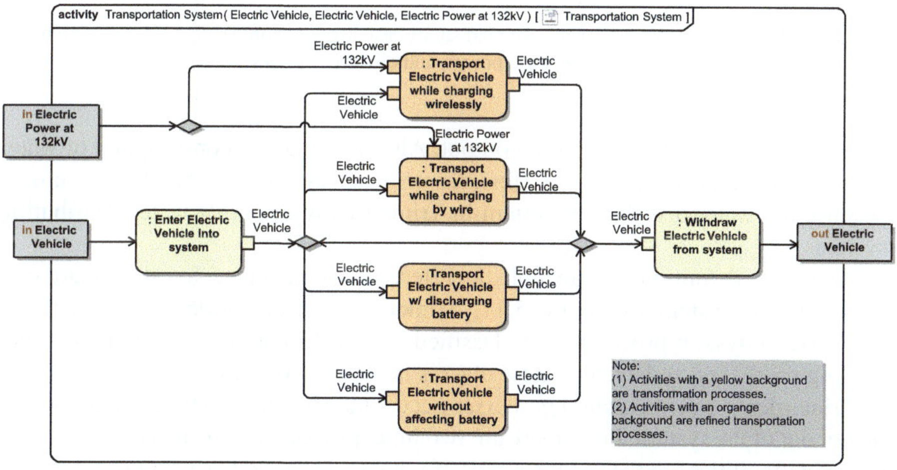

Fig. 5.10 Activity diagram of the electrified transportation system reference architecture

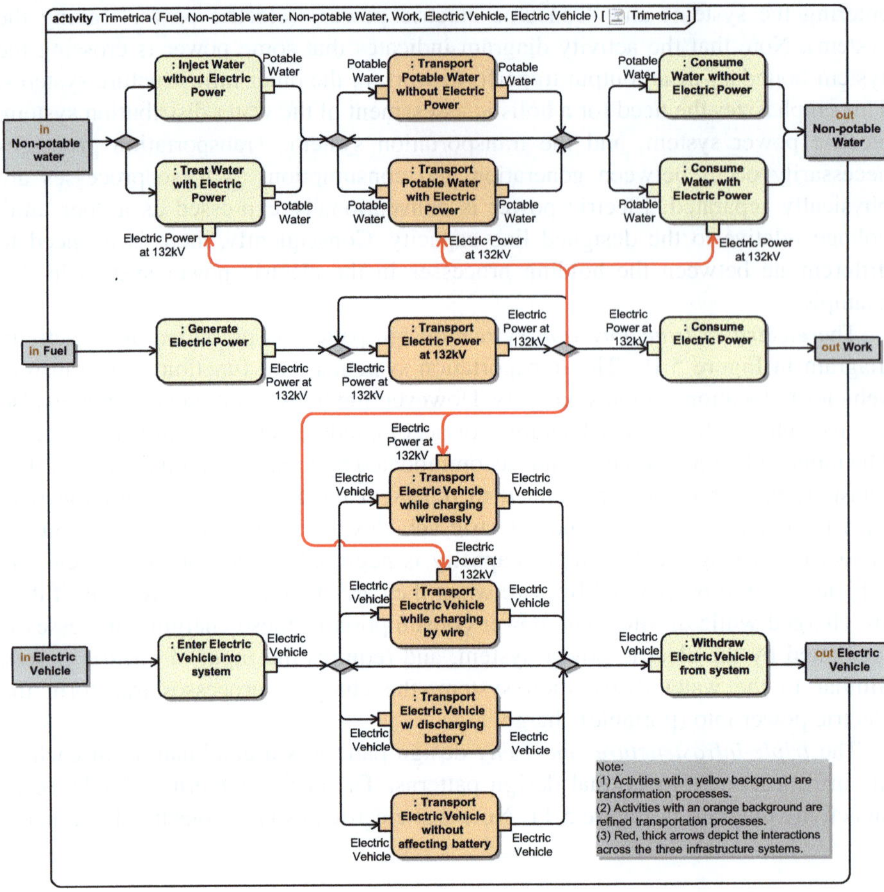

Fig. 5.11 Activity diagram of a triple operand smart city infrastructure system reference architecture: the three operands are water, electric power, and electric vehicles

whereas Figure 5.9 showed it as an output. Also note that the consumption of water by the electric power generator (for reasons such as cooling) has been omitted, because the water distribution system in Trimetrica is a *potable* water distribution system.

After the definition of the holistic reference architecture for the smart city infrastructure system in Figure 5.11, Trimetrica's instantiation can be derived. Trimetrica's system processes are classified as transformation, transportation, and holding processes, and each of these classes is now discussed in turn.

Trimetrica's *transformation processes* occur at the locations where operands flow into or out of the system. First, there are two input processes in Trimetrica. The water treatment facilities are the input sources of potable water, and use the process "treat water" to inject potable water across the system boundary. The input of electric power is performed by a single resource with a single "generate electric power" process. The transportation system is different in that it is assumed to be a closed system, without electric vehicles entering or leaving the system. As a result, no transformation processes are required for Trimetrica's transportation system.

The output of the system is determined by the consumption patterns of the distributed consumption resources. Water is consumed in houses and offices, with two different consumption types: consume hot water, and consume cold water. The water consumption processes are, for simplicity, assumed to be equal across the locations, resulting in a total of 2 transformation processes. The electric power consumption processes are also assumed to be equal across all consumption locations, and is thus added as a single transformation process. In conclusion, Trimetrica's set of system transformation processes is defined as $P_{\mu SC}$ ={treat water, generate electric power, consume hot water, consume cold water, consume electric power}.

Trimetrica's *refined transportation processes* require a discussion of both the holding processes and the transportation processes. Trimetrica's holding processes are derived following the three conditions that distinguish between transportation processes that:

- Have *different* operands,
- Hold a given operand in a *given* way.
- Change the *state* of the operand.

Trimetrica consists of three infrastructure systems with three separate operands. Consequently, the holding processes need to differentiate between those operands. The holding processes do not need to differentiate for the different ways to "hold" the operand in a given way. However, both the water distribution system and the transportation system differentiate the ways in which the state of the operand is changed. For the water distribution system, the water storage processes are assumed to consume electric power to, for example, pump the water to a water storage facility. Consequently, there are two holding processes for water: (1) carry potable water, and (2) carry potable water while consuming electric power.

The transportation system's operands are electric vehicles. There are four different types of movement of electric vehicles in Trimetrica [2, 9, 11, 20]. First, the vehicles can be stationary; without anything really changing. This is generally referred to as "parking." Second, the vehicles can be driving in which case the battery of the vehicle will discharge. Third, the vehicle can be charged while it is stationary. Usually, this is performed by wire for the fastest charging rates. Last, the vehicle can be charged while it is driving via induction (wireless) charging. Each of these four processes has a different impact on the state of the electric vehicle, and should consequently be included as holding processes. In conclusion, Trimetrica differentiates seven holding processes $P_{\gamma SC}$ = {carry potable water, carry potable water while consuming electric power, carry electric power at 132 kV, carry electric vehicle without affecting battery, carry electric vehicle while discharging, carry electric vehicle while charging by wire, carry electric vehicle while charging wirelessly}.

The transportation processes in Trimetrica are derived from the number of buffers in the smart city infrastructure system (see Equation 4.2). The number of unique buffers $\sigma(B_{SSC})$ equals 317, and the total number of transportation processes $P_{\eta SC}$ equals $\sigma(B_{SSC})^2$ = 100,489. The refined transportation processes $P_{\bar{\eta}SC}$ can be derived rather straightforwardly as the combination of each of the holding processes with all transportation processes: $\sigma(P_{\gamma SC})\sigma(P_{\eta SC}) = 703{,}423$. In summary, the Trimetrica test case has the following sets of processes:

- The set of Transformation Processes: $P_{\mu SC}$ has a size of 5.
- The set of Transportation Processes: $P_{\eta SC}$ has a size of 100,489.
- The set of Holding Processes: $P_{\gamma SC}$ has a size of 7.
- The set of Refined Transportation Processes: $P_{\bar{\eta}SC}$ has a size of 703,423.
- The set of System Processes: P_{SC} has a size of 703,428.

5.3.3 *Smart City Knowledge Base*

After the discussion of system form and system function, the section now uses the system knowledge base to map system function onto system form. Figure 5.12 introduces the SysML representation of system concept, by mapping the functions onto the instantiated resource classes.[7] For the discussion of system concept, the text will, however, refer to the atomic resources sets rather than the instantiated resource classes.

In Hetero-functional graph theory, the mapping of system function onto system form is conceived using the design equation, as introduced in Equation 4.6 on Page 48:

$$P = J_S \odot R \tag{5.1}$$

The Trimetrica knowledge base is calculated using the transformation knowledge base J_M, and the refined transportation knowledge base $J_{\bar{H}}$, following Equation 4.16 on Page 50:

$$J_S = \left[\begin{array}{c|c} J_M & \mathbf{0} \\ \hline \multicolumn{2}{c}{J_{\bar{H}}} \end{array}\right] \tag{5.2}$$

First, this section derives the *Trimetrica Transformation Knowledge Base*. The system transformation knowledge base was first introduced in Equation 4.11 on Page 49:

$$P_\mu = J_M \odot M \tag{5.3}$$

For Trimetrica, the set of transformation processes $P_{\mu SC}$ has a size of five. The set of transformation resources M_{SC} has a size of 201. The size of matrix J_{MSC}

[7]Note that Figure 5.12 includes the refined transportation processes rather than differentiating between regular transportation processes and holding processes.

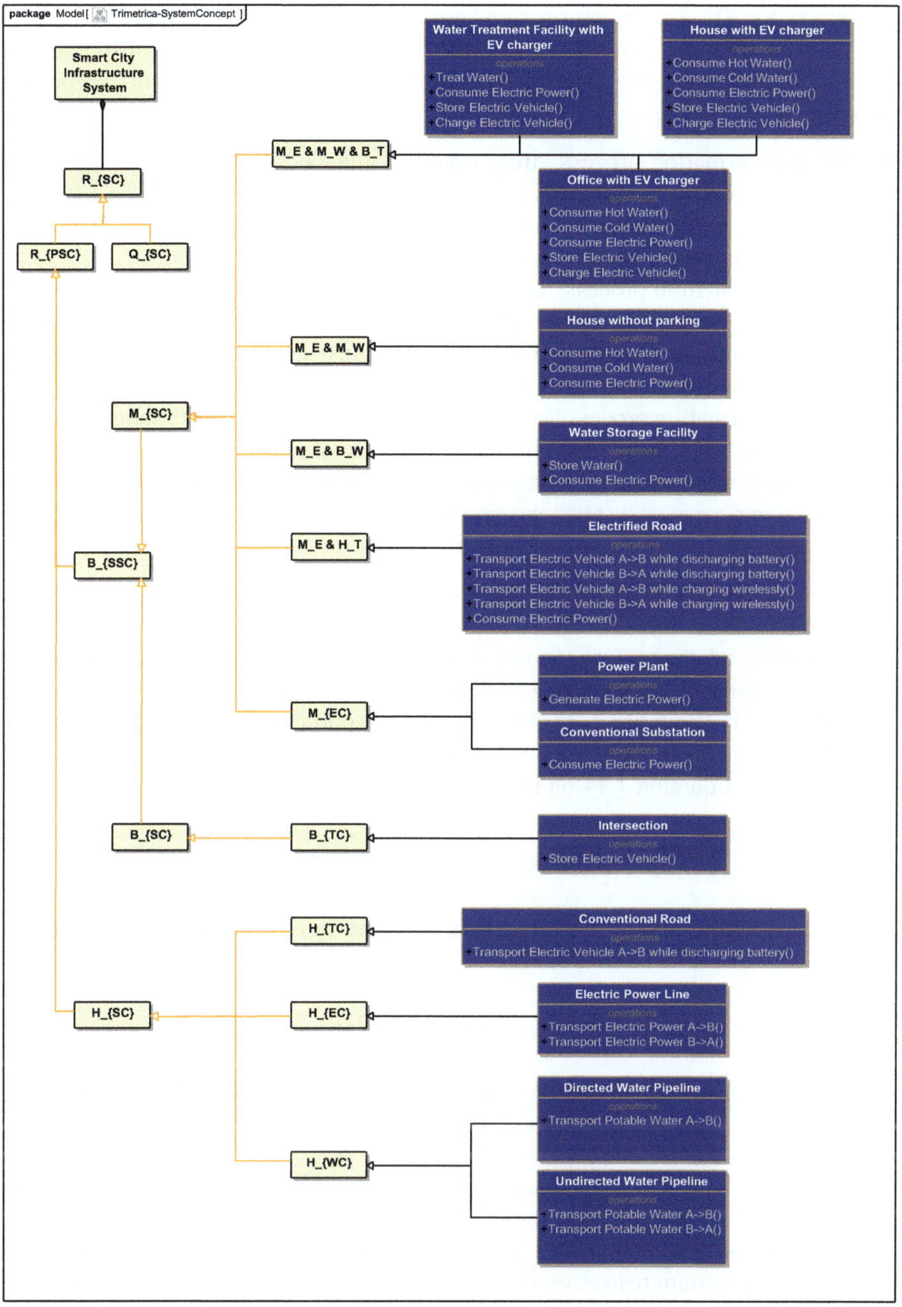

Fig. 5.12 SysML representation of the system concept for the Trimetrica smart city infrastructure system: this figure contains the unique set of capabilities for each of the resource classes in Trimetrica

follows as 5×201. The transformation knowledge base contains a degree of freedom for each mapping of a transformation process and a transformation resource. For each of the five transformation processes, the number of degrees of freedom is determined:

- The transformation process "treat water" is mapped onto five water treatment facilities, creating five degrees of freedom.
- The transformation process "generate electric power" is mapped onto the power plant, creating one degree of freedom.
- The transformation process "consume hot water" is mapped onto three resource classes: (1) house with EV charger, (2) office with EV charger, and (3) house without parking, of which Trimetrica contains 43, 5, and 62, respectively. Consume hot water, therefore, maps 110 times on to resources.
- The transformation process "consume cold water" maps onto the same resources as "consume hot water" and also has 110 degrees of freedom associated with it.
- The transformation process "consume electric power" is mapped onto all nodes in the electric power system, except the power plant. The total number of degrees of freedom related to "consume electric power" is thus 200.

In conclusion, the number of transformation degrees of freedom is: $5+1+110+110+200=426$.

Trimetrica's Refined Transportation Knowledge Base provides the mapping of the refined transportation processes $P_{\bar{\eta}}$ onto the system resources R_{SC}, using the design equation (Equation 5.1). The refined transportation knowledge base can be calculated using the holding knowledge base $J_{\gamma SC}$ and the transportation knowledge base J_{η}, using Equation 4.14 on Page 49:

$$J_{\bar{H}SC} = \left[J_{\gamma SC} \otimes \mathbb{1}^{\sigma(P_{\eta SC})}\right] \cdot \left[\mathbb{1}^{\sigma(P_{\gamma SC})} \otimes J_{HSC}\right] \tag{5.4}$$

where the transportation knowledge base J_{HSC} is derived using Equation 4.12 on Page 49:

$$P_{\eta SC} = J_{HSC} \odot R_{SC} \tag{5.5}$$

and where the holding knowledge base $J_{\gamma SC}$ is derived using Equation 4.13 on Page 49:

$$P_{\gamma SC} = J_{\gamma SC} \odot R_{SC} \tag{5.6}$$

The transportation knowledge base J_{HSC} maps the transportation processes $P_{\eta SC}$ onto the system resources R_{SC}, and has size $\sigma(P_{\eta SC}) \times \sigma(R_{SC})$ (or 100,489 $\times$ 1223). As a result, the transportation degrees of freedom can be divided into two sets: transportation degrees of freedom realized by transportation resources, and the transportation degrees of freedom realized by buffers. The former set of degrees of freedom is derived by matching the origin and destination of a transportation process to the physical origin and destination of the transportation resource. The elements in the set of conventional roads H_{TC} have a dedicated direction and perform a single transportation process. Their total number is 520, and consequently they enable 520 degrees of freedom. The electric power lines (in set H_{EC}), however, allow for a flow of power in either direction, and contain two degrees of freedom per resource. Their total number is thus 400, as each power line allows for a transportation process from its origin to the destination and back. The water pipelines (in H_{WC}) have 167 pipelines with direction, and 19 without. Consequently, their total degrees of freedom is $167 + 2(19) = 205$. For all of Trimetrica's transportation resources H_{SC}, the total number of transportation degrees of freedom is $520 + 400 + 205 = 1125$.

Trimetrica's set of buffers B_{SSC} consist of atomic resource classes that have transportation processes mapped to them. The majority of buffers in Trimetrica have a single location, and do not transport the operands between two different locations. Those sets are: $M_E\&M_W\&B_T$, $M_E\&M_W$, $M_E\&B_W$, M_{EC}, and B_{TC}. From these sets, only $M_E\&M_W$ and M_{EC} do not have the capability to store their respective operands, and therefore don't have a transportation process mapping to them. $M_E\&M_W\&B_T$, $M_E\&B_W$, and B_{TC} all have the ability to store an operand, or transport an operand from origin to destination where the origin equals the destination. For all elements in those three sets, the transportation knowledge base J_{HSC} contains a degree of freedom, which totals $53 + 10 + 116 = 179$.

There is, however, one set that does transport operands and is still considered a buffer: the set of electrified roads $M_E\&H_T$. This set has an origin and a destination, and allows one type of operand to be transported using the resource (electric vehicles). However, electric power is transformed into charge while the electric vehicles are transported. Consequently, the electrified road has a transformative nature and should be considered as a machine. After all, the classification "machine" supersedes the other classifications. The set of electrified roads, therefore, add 104 transportation degrees of freedom to the transportation knowledge base. In conclusion, the total number of Trimetrica's transportation degrees of freedom is $1125 + 179 + 104 = 1408$.

The holding knowledge base $J_{\gamma SC}$ maps the holding processes $P_{\gamma SC}$ onto the system resources R_{SC}, and has size $\sigma(P_{\gamma SC}) \times \sigma(R_{SC})$ (or 7×1223). As introduced previously, the holding processes are used to differentiate transportation processes in operand type, way of holding, or potential transformative nature. Each resource has at least one associated holding degree of freedom. The first set is: $M_E\&M_W\&B_T$. For each of the resources in this set, the resources are able to use potable water, electric power, and electric vehicles. Additionally, the electric vehicles can be charged, or parked while being held. Consequently, this set has four degrees of freedom per element, which totals 53*4 = 212 holding degrees of freedom. The set $M_E\&M_W$ is related to the operands water and electric power, and consequently has two associated degrees of freedom per element, which totals 62*2 = 124 holding degrees of freedom. The set $M_E\&B_W$ is related to the operands water and electric power, and consequently has two associated degrees of freedom per element, which totals 10*2 = 20 holding degrees of freedom. The set $M_E\&H_T$ is related to the operands electric power and electric vehicles, where the latter can be both charged wirelessly and discharged. Consequently, the resources have three associated degrees of freedom per element, which totals 52*3 = 156 holding degrees of freedom. The remaining sets M_{EC}, B_{TC}, H_{TC}, H_{EC}, and H_{WC} are all related to a single operand, which can only be held in a single way, and their transportation processes do not change the state of the operand. Consequently, the resources in these sets have a single holding process, with a total of $24+116+520+200+167+19=$ 1046. Based on the holding degrees of freedom for all the atomic resource sets, the total number of Trimetrica's holding degrees of freedom is $212+124+20+156+1046=$ 1558.

The refined transportation knowledge base $J_{\bar{H}SC}$ now combines the transportation knowledge base and the holding knowledge base using Equation 5.4. The resulting set of refined transportation processes are mapped onto the set of resources. Figure 5.12 presents the mapping of the refined transportation processes onto the Trimetrica's sets of resources. The resulting matrix has size $\sigma(P_{\bar{\eta}SC}) \times \sigma(R_{SC}) =$ 703,423 × 1223. The total number of refined transportation degrees of freedom is calculated using Equation 4.22 on Page 52:

$$\text{DOF}_{HSC} = \sum_{u}^{\sigma(P_{\bar{\eta}SC})} \sum_{v}^{\sigma(R_{SC})} \left[J_{\bar{H}SC} \ominus K_{\bar{H}SC}\right](u, v) \tag{5.7}$$

where K_{HSC} is all zeros. The total number of refined transportation degrees of freedom DOF_{HSC} is 1565.

In conclusion, the Trimetrica knowledge base J_{SSC}, constraints matrix K_{SSC}, and system concept A_{SSC} have size 703,428 × 1223. The total number of degrees of freedom can be calculated using Equation 4.18 on Page 52, or adjusted for the Trimetrica test case:

$$\text{DOF}_{SSC} = \sigma(\mathcal{E}_S) = \sum_{w}^{\sigma(P_{SC})} \sum_{v}^{\sigma(R_{SC})} [J_{SSC} \ominus K_{SSC}](w, v) \tag{5.8}$$

which equals 1991.

5.3.4 *Visualizing Degrees of Freedom*

In order to intuitively understand the meaning of Trimetrica's set of degrees of freedom, a visualization is required. The degrees of freedom are a mapping of system function onto system form, and therefore, a mapping of system processes onto system resources. Each of those mappings is a degree of freedom, or a capability of the interdependent infrastructure system. Logically, these system capabilities can be represented per resource while adopting the physical location of their associated resource.

Trimetrica is originally depicted in Figure 5.1 on Page 97. This section compares each of the three separate infrastructure systems to the degrees of freedom associated with these systems. Figure 5.13 provides a comparison between the Trimetrica water distribution system topology and its degrees of freedom. Three observations are made. First, the water treatment facilities have a single capability. This capability acts on two operands, because it consumes electric power to generate potable water. Second, all water consumption nodes, whether they are houses or offices, have two associated degrees of freedom. These represent both hot water consumption and cold water consumption where the former has two operands, and the latter only consumes potable water as its operand. Lastly, the capabilities of the water pipes are displayed in between the end-points of the water pipe. Whenever the end-points are more than two units of length away, the capability is located at 75% of the pipe's length. Otherwise, the capability is located in the middle. The capabilities associated with the water pipes have potable water as a single operand.

Figure 5.14 provides a comparison between the Trimetrica electric power system topology and its degrees of freedom. Three observations are made. First, there is overlap between Figures 5.13 and 5.14, because the hot water consumption, water treatment, and water storage processes all consume electric power, and transform potable water. Second, all power transmission lines have two associated degrees of freedom to represent the potential two-way flow of electric power. Lastly, the figure contains overlap with Figure 5.15, because it includes the charging capabilities of the transportation system; whether they are charging roads or conventional charge-by-wire facilities.

Figure 5.15 provides a comparison between the Trimetrica electrified transportation system topology and its degrees of freedom. Two observations are made. First, the charging capabilities in pink coincide with either two conventional transportation capabilities or a single parking capability. There are no capabilities where charging is the sole capability. Second, the intersections have a "storage" capability. It is, however, zero, because parking a vehicle on an intersection is not allowed. It, therefore, represents the connection between two or more roads.

Figure 5.16 superimposes the degrees of freedom of the five operands in Trimetrica. The figure contains a window in the bottom left corner. The enlarged depiction of this window is presented in Figure 5.17, with the topological presentation of the same area. Figure 5.17 presents the degrees of freedom provided by the topology of the six resources. The House 1 (or resource 6) has five associated degrees of freedom:

1. consume cold potable water at house 1
2. consume hot potable water at house 1
3. consume electric power at house 1
4. park electric vehicle at house 1
5. charge electric vehicle by wire at house 1

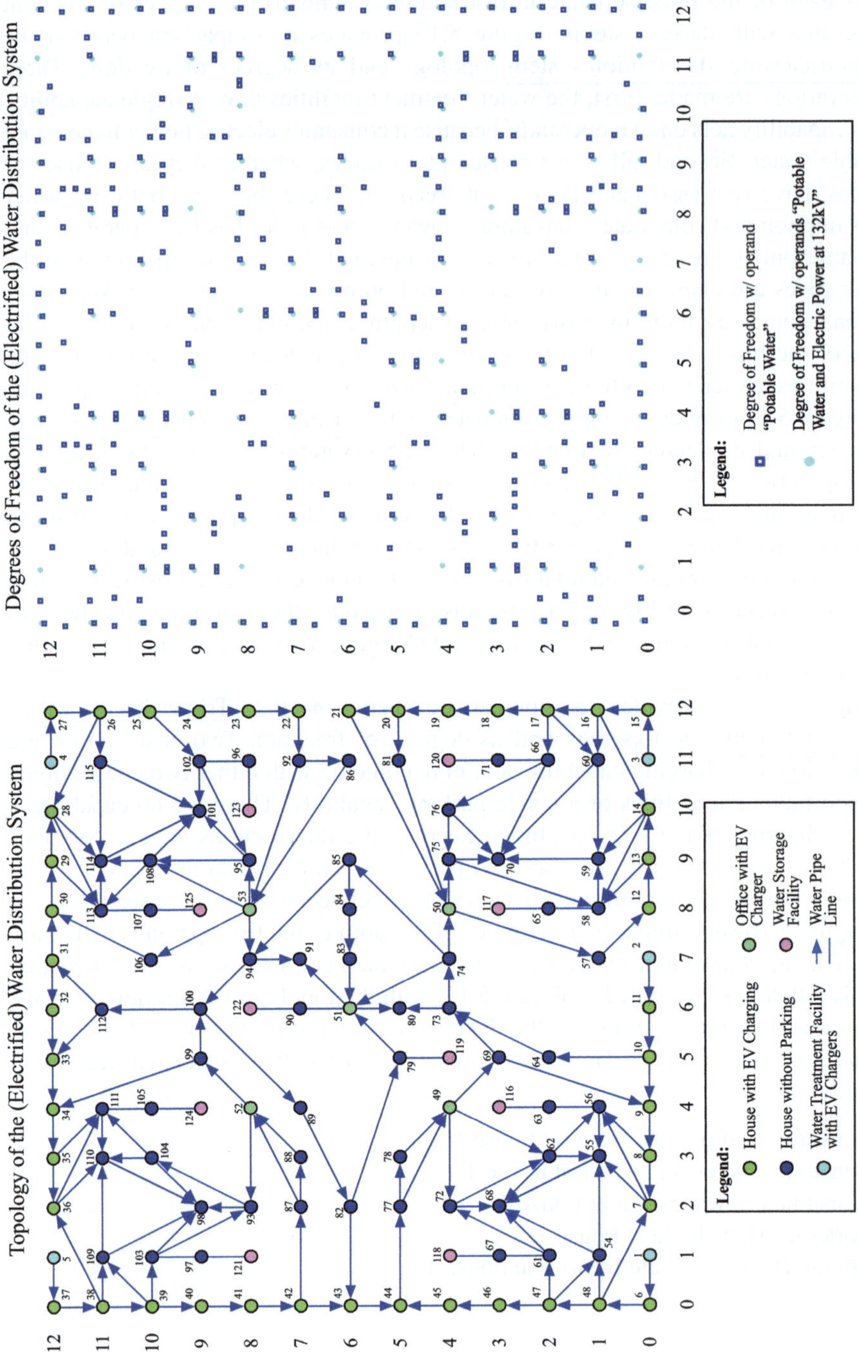

Fig. 5.13 Structural degrees of freedom of water-related operands

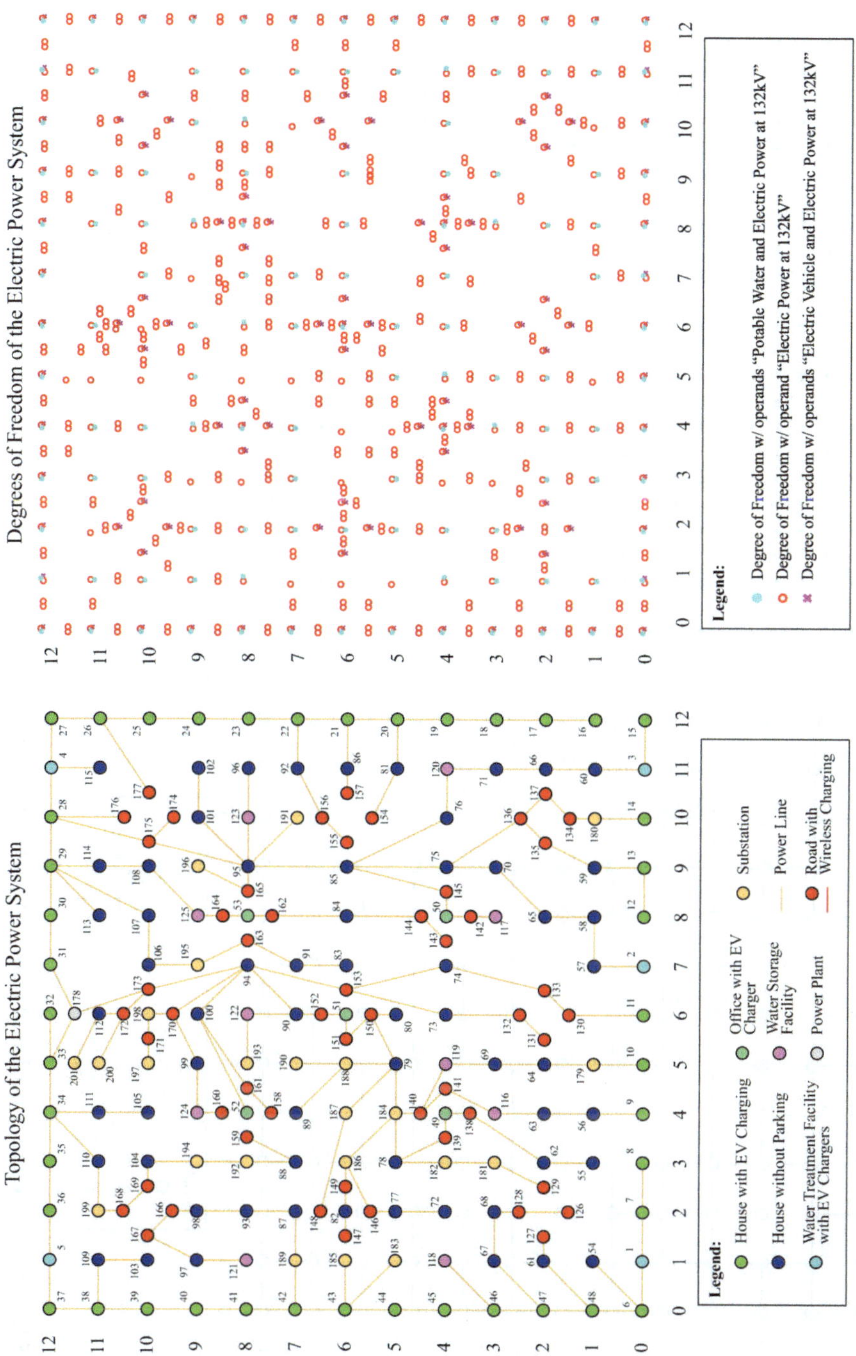

Fig. 5.14 Structural degrees of freedom of electricity-related operands

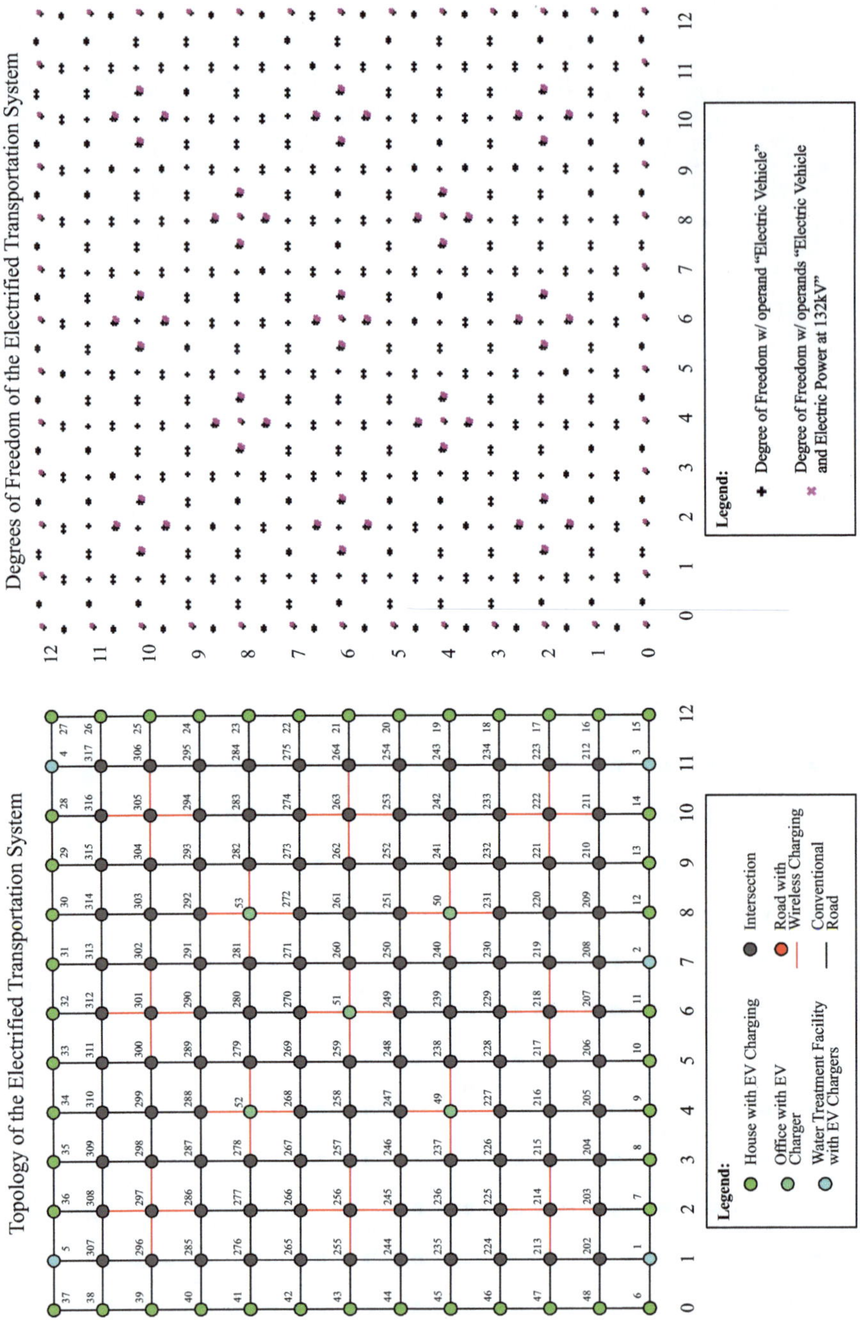

Fig. 5.15 Structural degrees of freedom of transportation-related operands

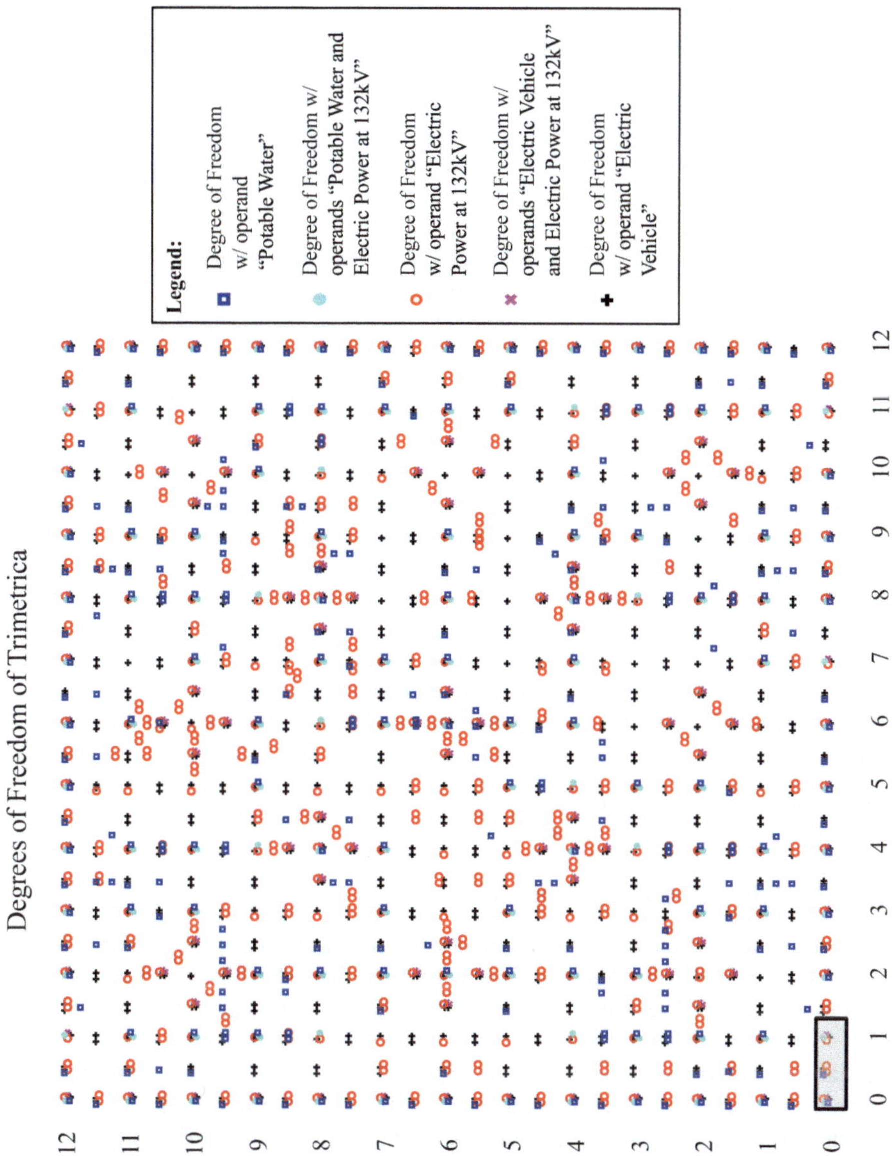

Fig. 5.16 Trimetrica's structural degrees of freedom: the frame in the bottom-left corner indicates the detail presented in Figure 5.17 on Page 124

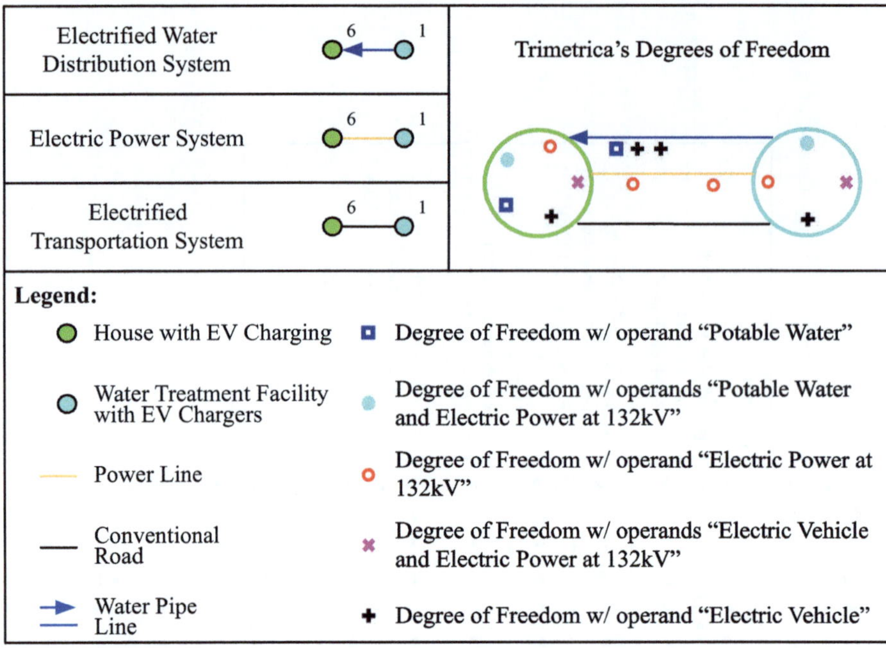

Fig. 5.17 A comparison between a detail of Trimetrica's topology and the same detail of Trimetrica's structural degrees of freedom as indicated in Figure 5.2 on Page 100 and Figure 5.16 on Page 123

All degrees of freedom are of a different operand type, and are depicted with their associated symbols. Water treatment facility 1 has four associated degrees of freedom:

1. treat water at water treatment facility 1
2. consume power at water treatment facility 1
3. park EV at water treatment facility 1
4. charge EV by wire at water treatment facility 1

Trimetrica contains four different transportation resources between house 1 and water treatment facility 1. The first is a water pipeline from the water treatment facility to the house. The water pipeline transports water from the water treatment facility to the house, and has, therefore, a single associated degree of freedom. Note that the test case assumes the water pipeline is loss-less and does not require additional pressurization. The second resource between the house and the water treatment facility is an electric power line. It facilitates transmission of electric power between the house and the water treatment facility in both directions. Consequently, there are two electric power degrees of freedom. Lastly, there are two roads, one originating in the house, and the other originating in the water treatment facility. Note that the roads are depicted as a single line. Each of the two roads has a single degree of freedom, and therefore, there are two degrees of freedom of operand class "electric vehicle."

In conclusion, this section first determined system form and system function for Trimetrica using the block definition diagram and activity diagram in SysML. Based on this conceptual framework, the set of system resources and processes were defined. Subsequently, the system knowledge base was calculated as a map of system function onto system form. Lastly, the section visualized the degrees of freedom by a comparison with topological lay-outs of Trimetrica's water distribution, electric power, and electrified transportation systems.

5.4 Hetero-functional Adjacency Matrix

Following the discussion of Trimetrica's degrees of freedom, that represent the capabilities of the smart city infrastructure system, this section continues to define their logical sequence. The section applies the theoretical foundation defined in Section 4.2 on Page 55 to the Trimetrica test case. First, the numerical results are discussed based on the application of the mathematical model. Then, the degree of freedom visualizations from the previous section are expanded to incorporate system sequence for a more intuitive understanding.

5.4.1 Calculating System Sequence

In the previous chapter, the Hetero-functional Adjacency Matrix (HFAM) A_ρ was first introduced as the Boolean difference between the system sequence knowledge base J_ρ and the system sequence constraints matrix K_ρ, using Equation 4.25. When applied to the Trimetrica test case, the equation is:

$$A_{\rho SC} = J_{\rho SC} \ominus K_{\rho SC} \tag{5.9}$$

where $J_{\rho SC}$ is the system sequence knowledge base, and where $K_{\rho SC}$ is the system sequence constraints matrix. The size of matrices $A_{\rho SC}$, $J_{\rho SC}$, and $K_{\rho SC}$ equals: $\sigma(R_{SC})\sigma(P_{SC}) \times \sigma(R_{SC})\sigma(P_{SC})$, or numerically: 860,292,444 × 860,292,444.

Note that the system sequence knowledge base J_ρ defines all combinations of degrees of freedom (as in Definition 4.14 on Page 55). When Equation 4.26 is applied to the Trimetrica test case, $J_{\rho SC}$ is calculated as follows:

$$J_{\rho SC} = A^{V}_{SSC} A^{VT}_{SSC} \tag{5.10}$$

where $J_{\rho SC}$ has $\text{DOF}^2_{SSC} = 1991^2 = 3{,}964{,}081$ filled elements.

Since not all combinations of degrees of freedom (DOFs) are feasible, the system sequence constraints matrix is introduced (as in Definition 4.15 on Page 56). The system sequence constraints matrix enforces two types of sequence constraints on the system sequence knowledge base. First, it imposes the physical continuity constraints presented in Table 4.3 on Page 57. Second, it imposes the functional sequence constraints which are based on the reference architecture (or design pattern) of the modeled system.

The system sequence constraints matrix is calculated using the aforementioned constraints. Even though this approach is correct, it is not always practical from a computational perspective, because it requires $\left[\sigma(R)\sigma(P)\right]^2 = 860{,}292{,}444^2$ calculations. Since the $J_{\rho SC}$ matrix is sparse, it is computationally more efficient to check each of the logical DOF sequences for compliance with the imposed constraints. The latter approach facilitates the computation of $A_{\rho SC}$, avoids the calculation of the memory-intensive $K_{\rho SC}$ matrix, and requires only $\text{DOF}_{SSC}^2 = 1991^2$ calculations. If the logical sequence of DOFs complies with the constraints, it is transferred to $A_{\rho SC}$. The calculation of the Hetero-functional Adjacency Matrix for the Trimetrica test case uses this latter approach. The number of DOF sequences that comply with each of the physical continuity constraints is listed below:

- $\text{DOF}_{R1} = 1096$
- $\text{DOF}_{R2} = 2309$
- $\text{DOF}_{R3} = 2314$
- $\text{DOF}_{R4} = 9902$

This reduces the total number of sequence degrees of freedom that adhere to the physical continuity constraints to 15,621.

The DOF sequences that comply with the functional sequence constraints are derived from the Trimetrica design pattern in Figure 5.11. The total number of DOF sequences that comply with the design pattern is 1,350,300.

The intersection of these two sets of DOF sequences are included in $A_{\rho SC}$. The resulting number of sequence degrees of freedom that comply with both the functional sequence and continuity constraints $\text{DOF}_{\rho SC}$ is 8082.

Note that the hetero-functional adjacency matrix can be projected to reduce its size to more manageable levels. Equation 4.31 on Page 58 demonstrates the projection of any hetero-functional adjacency matrix. When applied to the Trimetrica test case, it follows:

$$\mathbb{P}_S A_{\rho SC} \mathbb{P}_S^T = \tilde{A}_{\rho SC} \tag{5.11}$$

The size of the projected matrix is $\sigma(\mathcal{E}_S) \times \sigma(\mathcal{E}_S)$. No information is lost in this process, and the number of sequence degrees of freedom remains naturally unchanged.

5.4.2 *Visualizing System Sequence*

Visualizations of the HFAM serve to facilitate its understanding. Figures 5.13, 5.14, and 5.15, which present DOFs as a collection of scattered nodes, are now revised with connecting arcs that represent system sequence. Note that although the adjacency matrix is a directed graph the figures in this section do not contain arrows for the sake of clarity. The detailed view of the hetero-functional adjacency matrix does, however, include explicitly directional sequences.

Figure 5.18 on page 128 compares the electrified water distribution system topology with the hetero-functional adjacency matrix that only water-related degrees of freedom. The visualization of the hetero-functional adjacency matrix closely resembles the topology of the water distribution system. The sequence degrees of freedom require physical continuity and, therefore, a physical interface between the resources is needed.

Figure 5.19 on page 129 compares the electric power system with the hetero-functional adjacency matrix, including only electric power related degrees of freedom. Other than the electrified water distribution system, the transportation resources each have two degrees of freedom to accommodate the bidirectional flow of electric power. Additionally, one can notice that all degrees of freedom are coupled, and consequently, all degrees of freedom are able to connect to the electric power system. This is essential, because the water consumption, treatment, and storage degrees of freedom, as well as the EV charging degrees of freedom rely on the power grid for their operation.

Figure 5.20 on page 130 compares the electrified transportation system with the hetero-functional adjacency matrix, including only transportation related degrees of freedom. Note that the degrees of freedom of the roads are connected in a diamond like pattern. It is caused by the sequences in which two transportation processes around the same intersection are coupled.

Based on the previous three figures, it becomes clear that the degrees of freedom are coupled across the operand types. Conventionally, in graph theory, the systems with different operand types have been separated into layers per operand [6]. As discussed in Chapter 2, this practice limits the way the operand-specific networks are coupled. Hetero-functional graph theory, however, allows the coupling of nodes across layers based on physical continuity and functional sequence. In the end, an interdependent infrastructure system is a single system; rather than a number of separate systems. Hetero-functional graph theory reflects this reality as a single network in which function, form, and operand type are uniquely defined. Figure 5.22 on page 132, now, presents the full hetero-functional adjacency matrix for Trimetrica. The complexity of the Trimetrica interdependent smart city infrastructure system creates a very crowded representation. Figure 5.21 on page 131 provides a detail of the Trimetrica hetero-functional adjacency matrix (Figure 5.22), in which 25 sequences of 14 degrees of freedom are provided. This figure builds on Figures 5.2 and 5.17 on pages 100 and 124, respectively.

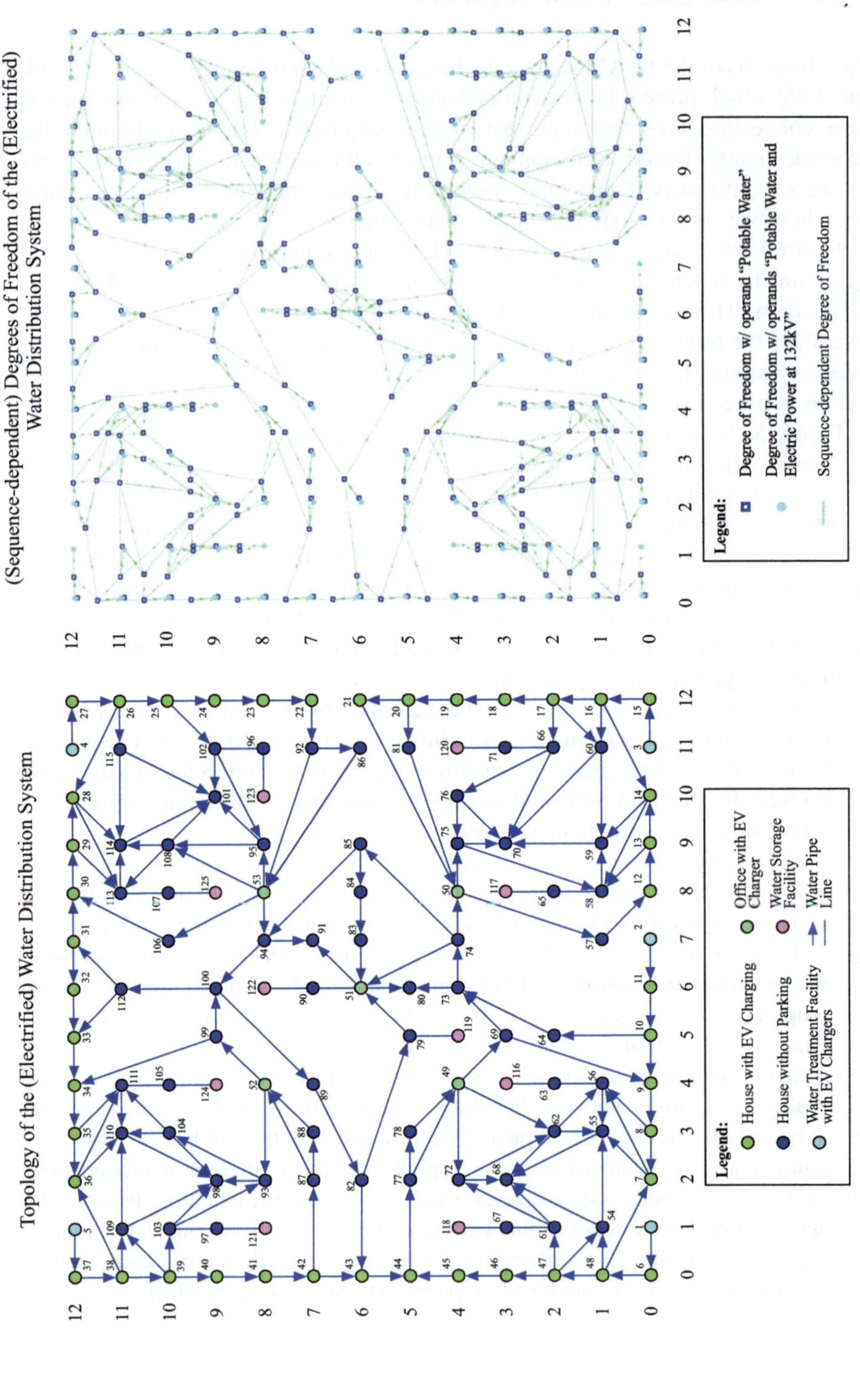

Fig. 5.18 Sequence-dependent degrees of freedom of the operands: water, and water with electric power at 132 kV

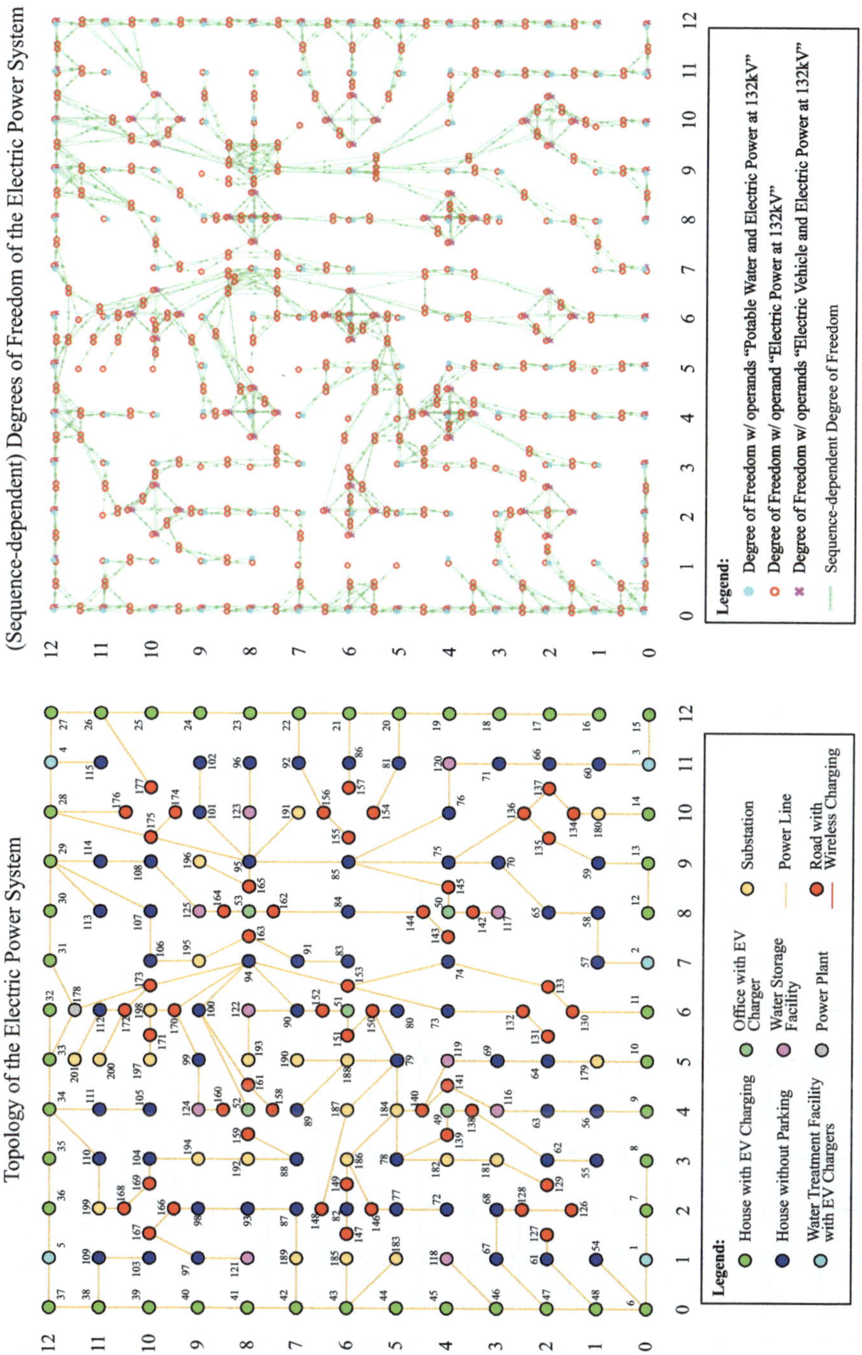

Fig. 5.19 Sequence-dependent degrees of freedom of the operands: electric power at 132 kV, water with electric power at 132 kV, and EV with electric power at 132 kV

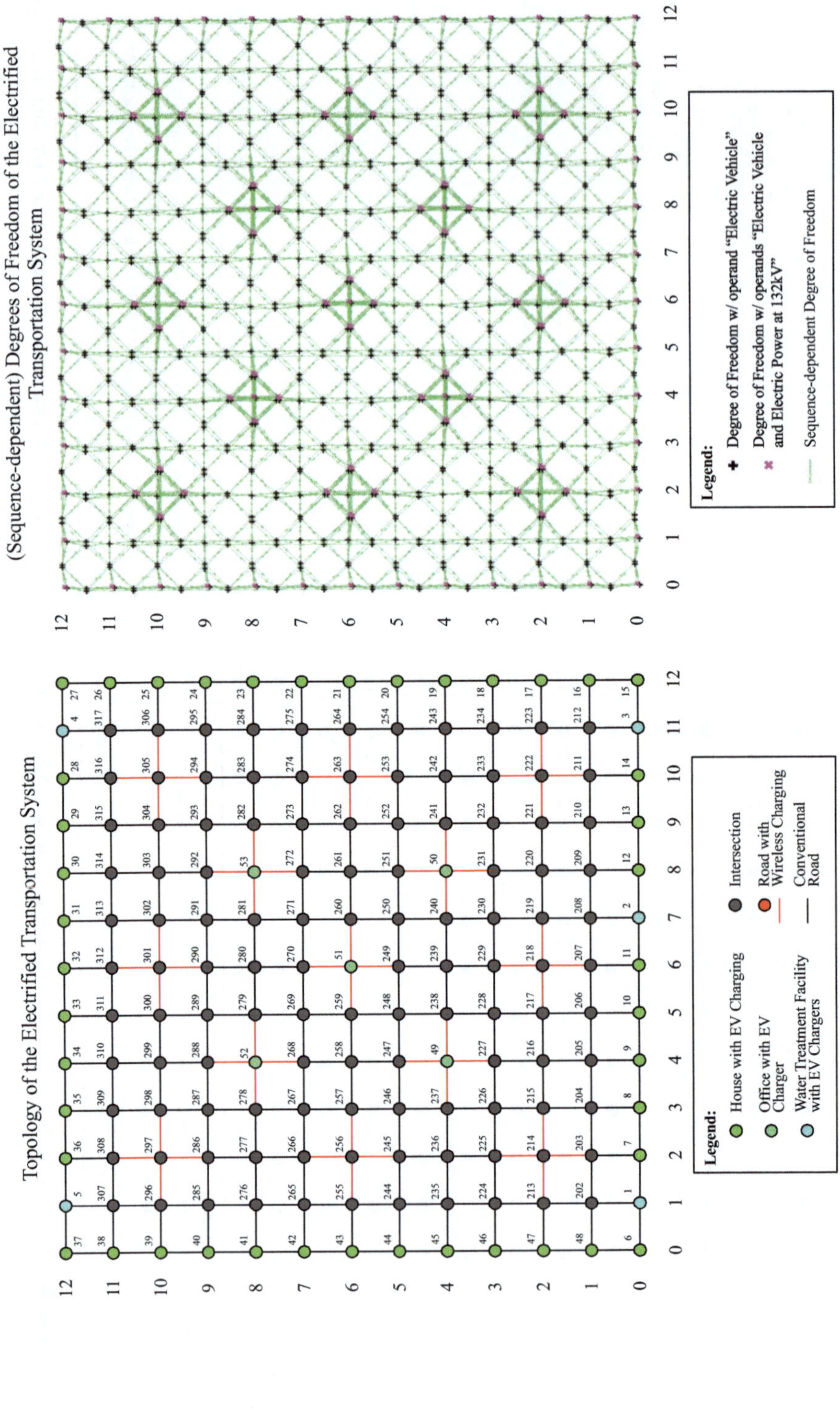

Fig. 5.20 Sequence-dependent degrees of freedom of the operands: EV, and EV with electric power at 132 kV

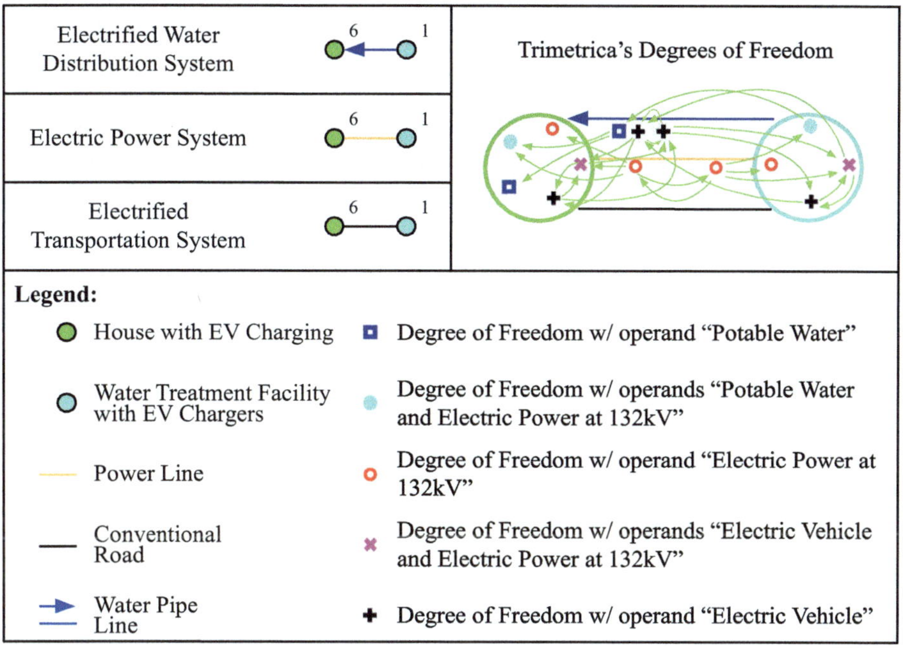

Fig. 5.21 A comparison between a detail of Trimetrica's topology and the same detail of Trimetrica's structural degrees of freedom as indicated in Figure 5.2 on Page 100 and Figure 5.16 on Page 123

Figure 5.23 on page 133 represents the hetero-functional adjacency matrix as a network with five distinct layers. Each of the layers contains a unique set of degrees of freedom classified by operand type. System sequence across as well as within all five operand classes is provided.

In conclusion, this section has provided an overview of the mathematical derivation of the hetero-functional adjacency matrix for the Trimetrica test case. First, the set of feasible sequence degrees of freedom was calculated and captured in the hetero-functional adjacency matrix. Then, the hetero-functional adjacency matrix for each of the infrastructure systems in Trimetrica was visualized as a means to compare the adjacency matrix to the physical topology. Lastly, all degrees of freedom were superimposed and coupled using the hetero-functional adjacency matrix. The hetero-functional adjacency matrix was used as a structural coupling of the three infrastructure layers with five different operand classes.

5.5 Controller Agency Matrix

The controller agency matrix is introduced in Section 4.3 on Page 60 as a means to differentiate between two systems of equivalent capabilities but different control structure. The Trimetrica test case contains a control structure that includes a water

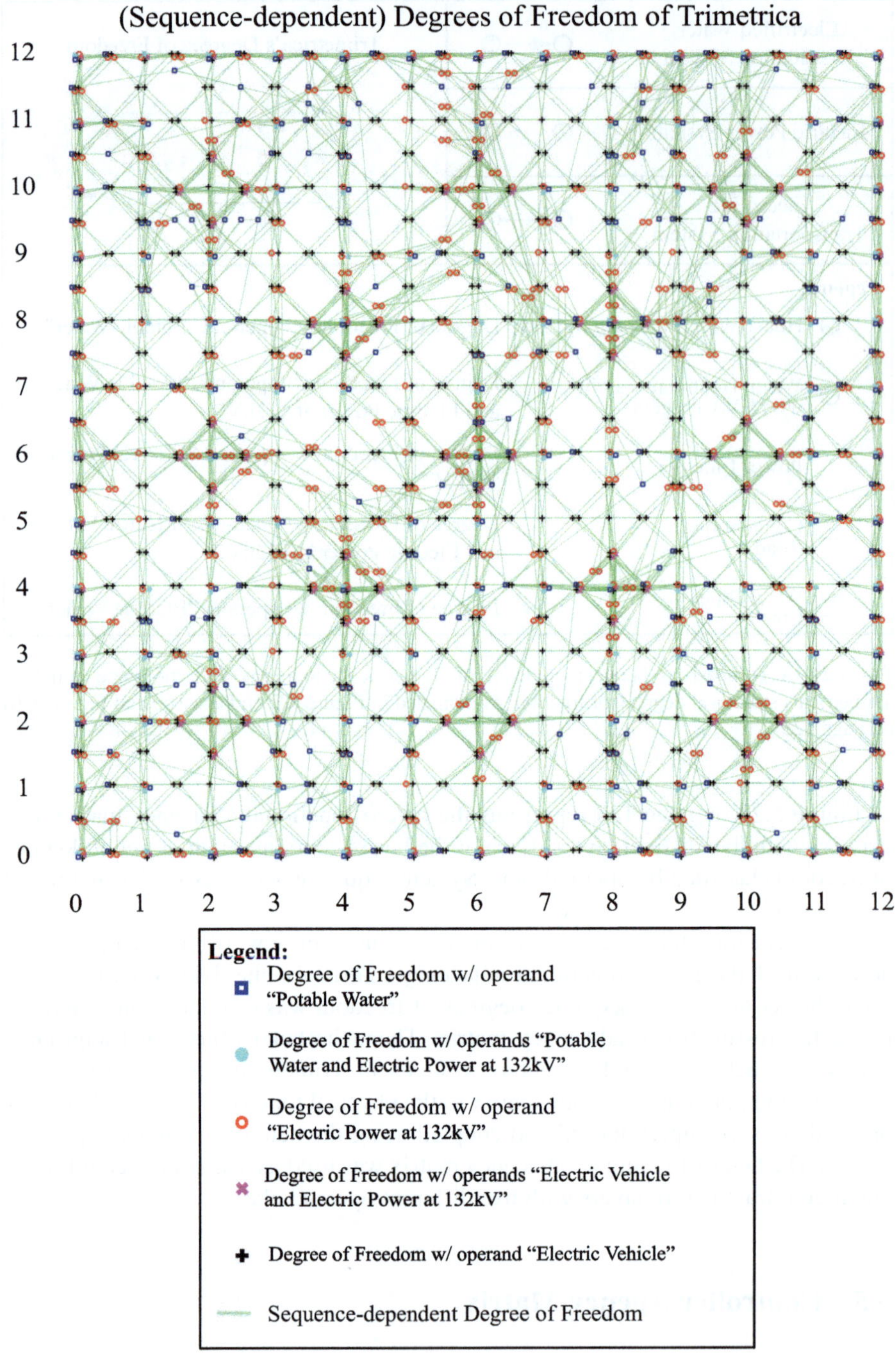

Fig. 5.22 Trimetrica's Hetero-functional adjacency matrix with all five layers of degrees of freedom in a single plane

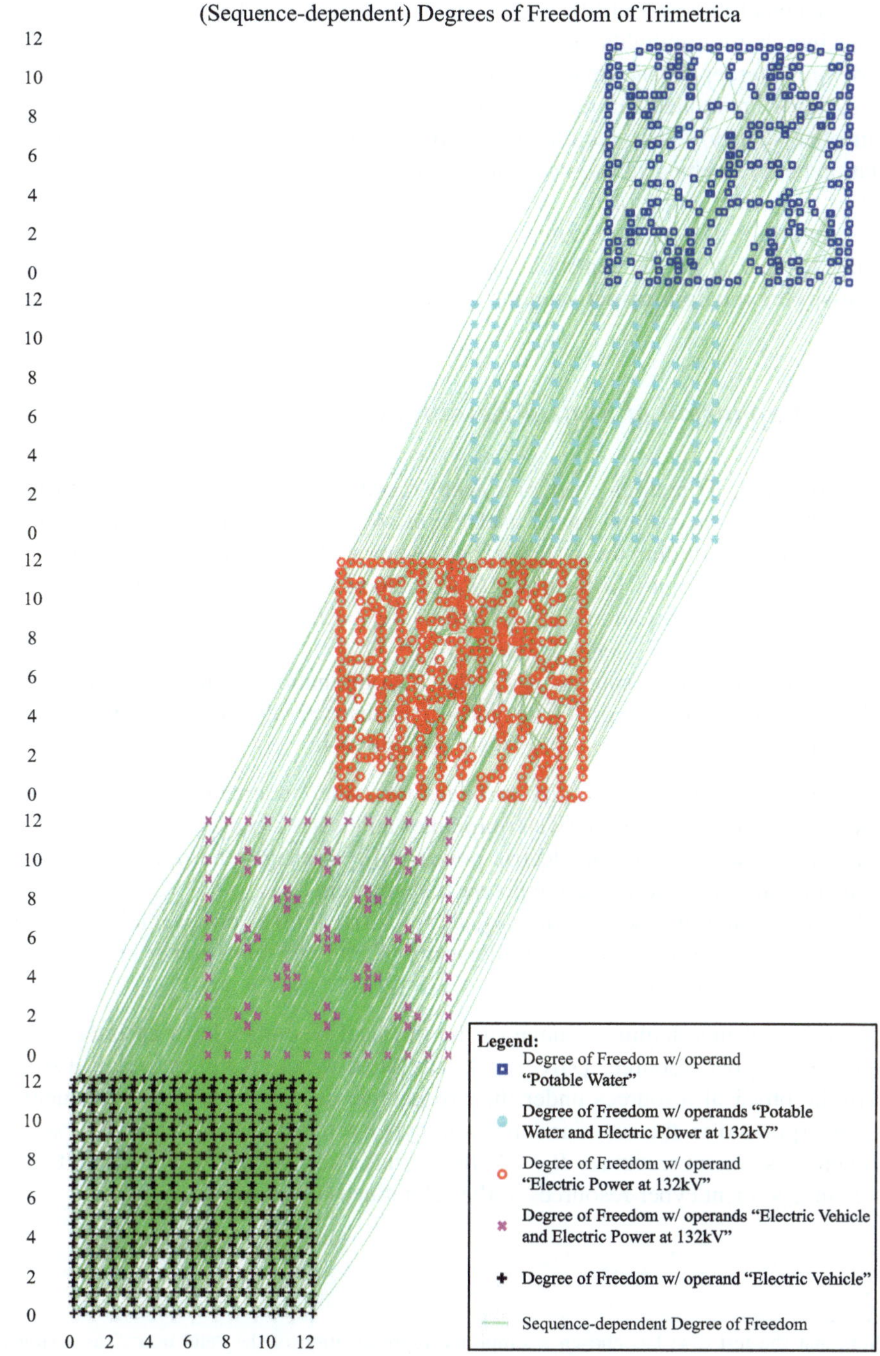

Fig. 5.23 Trimetrica's hetero-functional adjacency matrix presented as a five-layer network

utility, an electric power utility, and an end user as its control agents. This example is succinct, but it highlights the importance of including the control structure when modeling interdependent smart city infrastructure systems. As a consequence of the systems' integrated nature, the legacy control structures have to cooperate. The controller agency matrix facilitates the holistic operation, because it clarifies the relation between physical resources and their controllers.

For the construction of the controller agency matrix, the section first expands the existing set of system resources to include Trimetrica's cyber-resources. Based on this expanded set, the controller agency matrix is calculated. Lastly, the section discusses the relationship between the cyber-resources and the degrees of freedom.

5.5.1 Expansion of System Resources

In Section 5.3, Trimetrica's set of system resources R_{SC} was defined with size 1223. However, the section based itself on the definition of the physical resources R_{PSC}, without including the system cyber-resources Q_{SC}. Following the expansion of the system resources in Section 4.3 to include the cyber-resources, the set of Trimetrica's system resources is redefined to $R_{SC} = R_{PSC} \cup Q_{SC} = M_{SC} \cup B_{SC} \cup H_{SC} \cup Q_{SC}$.

Trimetrica's cyber-resources Q_{SC} are a combination of the dependent and independent cyber-resources, Q_{DSC} and Q_{ISC}, respectively. For the Trimetrica test case, three types of cyber-resources have been defined:

1. Water Utility: the water utility manages the water treatment facilities and the water distribution network (water pipelines and water storage facilities).
2. Electric Power Utility: the electric power utility manages the power generation facility, the substations and the power transmission lines.
3. End Users: the end users are individuals who drive their electric vehicles on roads (with or without charging) and intersections, and consume power and water at home or in the office.[8]

As a result of this definition, none of the physical resources have an internal (or dependent) cyber-resource. $Q_{DSC} = \emptyset$. The three cyber-resources are all distinct from the physical resources under their respective jurisdictions, and consequently are incorporated in the set of independent cyber-resources Q_{ISC}. The final set of Trimetrica's system resources R_{SC}, therefore, has a size of $1223+3 = 1226$. It adds three independent cyber-resources to the set defined in Section 5.3.

[8]Note that the test case has chosen a simplified representation of the "end users," as a single aggregated cyber-resource. One can also decide to represent each of the end users as a separate cyber-resource. However, the current representation has been chosen to reduce the complexity of the visualizations and maintain intuition.

5.5.2 Smart City Controller Agency Matrix

The controller agency matrix is defined in Definition 4.18 on Page 62 as a binary matrix of size $\sigma(R_P) \times \sigma(R)$, whose element $A(v_1, v_2)$ is equal to one when the resource $r_{v_2} \in R$ has control jurisdiction over the physical resource $r_{v_1} \in R_P$. For Trimetrica, the resulting controller agency matrix A_{QSC} has a size of 1223 × 1226 with 1223 filled elements. Each physical resource is under the jurisdiction of a single independent cyber-resource.

$$A_{QSC} = \left[I^{\sigma(R_P)} \mid \overline{A}_{QSC} \right] \tag{5.12}$$

As a result of this characteristic, the matrix can be reduced to $\overline{A}_{QSC}$ with a size of $\sigma(R_{PSC}) \times \sigma(Q_{ISC}) = 1223 \times 3$.

5.5.3 The Relation Between the Controller Agency Matrix and the Hetero-functional Adjacency Matrix

The previous two sections defined Trimetrica's interdependent smart city infrastructure system from a structural perspective using degrees of freedom and sequence-dependent degrees of freedom to create the Hetero-functional Adjacency Matrix. The structural degrees of freedom represent a mapping of the system processes onto the system resources. The controller agency matrix defines the control relation between the cyber-resources and the physical resources. Consequently, the controller agency matrix implies a coupling of the physical resources' degrees of freedom. The controller agency matrix can, therefore, be expanded to couple the controller agents to the hetero-functional adjacency matrix, with matrix $\widehat{A}_{QSC}$:

$$\widehat{A}_{QSC} = \overline{A}_{QSC} \otimes \mathbb{1}^{\sigma(P)} \tag{5.13}$$

where $\widehat{A}_{QSC}$ has size $\sigma(R)\sigma(P) \times \sigma(Q)$ Additionally, this matrix can be projected to create $\widetilde{A}_{QSC}$, so that controller agents are coupled to the projected set of capabilities.

$$\widetilde{A}_{QSC} = \mathbb{P}_S(\overline{A}_{QSC} \otimes \mathbb{1}^{\sigma(P)}) \tag{5.14}$$

where $\widetilde{A}_{QSC}$ has size $\mathcal{E}_S \times \sigma(Q_{SC}) = 1991 \times 3$ with 1991 filled elements.

Conventionally, jurisdiction has been imposed upon physical resources without consideration of their functions. For example, the water utility controls not just the water treatment facility but also parking permits. Therefore, it determines who is allowed to park and charge at the water treatment facility's parking lot. Consequently, the system degrees of freedom associated with transportation or electric power may be uncoordinated with those infrastructures.

The controller agency matrix is represented in Figure 5.24 on Page 137 as a coupling between cyber-resources and system degrees of freedom. Combining this coupling (found in Figure 5.24 on Page 137) with the hetero-functional adjacency matrix in Figure 5.23 on Page 133 yields an integrated picture of cyber-physical couplings in Figure 5.25 on Page 138.

5.6 Controller Adjacency Matrix

Section 4.4 on Page 64 introduced the Controller Adjacency Matrix as the third of three types of interfaces identified by Hetero-functional Graph Theory. The first interface type is captured by the Hetero-functional Adjacency Matrix, and addresses the interface between physical capabilities. The second interface is captured by the Controller Agency Matrix and addresses interface between the physical resources and cyber-resources. The Controller Adjacency Matrix captures the third and final interface and addresses the cyber-interface between the cyber-resources. The controller adjacency matrix in Definition 4.19 is a binary matrix of size $\sigma(Q) \times \sigma(Q)$. Its elements $A_C(v_1, v_2)$ are equal to one when cyber-resource $r_{v_1} \in Q$ passes information to cyber-resource $r_{v_2} \in Q$.

Trimetrica's set of cyber-resources $Q_{SC} = \{$water utility, electric power utility, end user$\}$ was introduced previously in Section 5.5. Consequently, the controller adjacency matrix A_{CSC} has a size of 3×3. The matrix captures all cyber-interfaces between the cyber-resources, and, therefore, all elements are connected.

- The End Users (physically) draw electric power and water from the respective systems. The physical exchange is facilitated by information about payments, location, and consumption.
- The Water Utility provides potable water to the end users. This physical exchange is enabled by a payment system, and water consumption measurement devices. Note that the water utility also draws electric power for its water treatment facilities. Therefore, there is also a direct cyber-interface between the water utility and the electric power utility.
- The Electric Power Utility supplies electric power to the end users and the water utility. The physical exchange is facilitated by information about electric power consumption patterns and a payment system.

As a result of these interactions, all 9 elements in Trimetrica's controller adjacency matrix are filled.

Figure 5.26 presents an overview of the three cyber-resources in the Trimetrica test case, with their respective cyber-interfaces. Logically, all three cyber-resources are directly connected.

Figure 5.27 on Page 140 superimposes the controller adjacency matrix on Figure 5.25 as introduced in Section 5.5. The figure now shows all three types of interfaces between physical, and cyber-resources in Trimetrica.

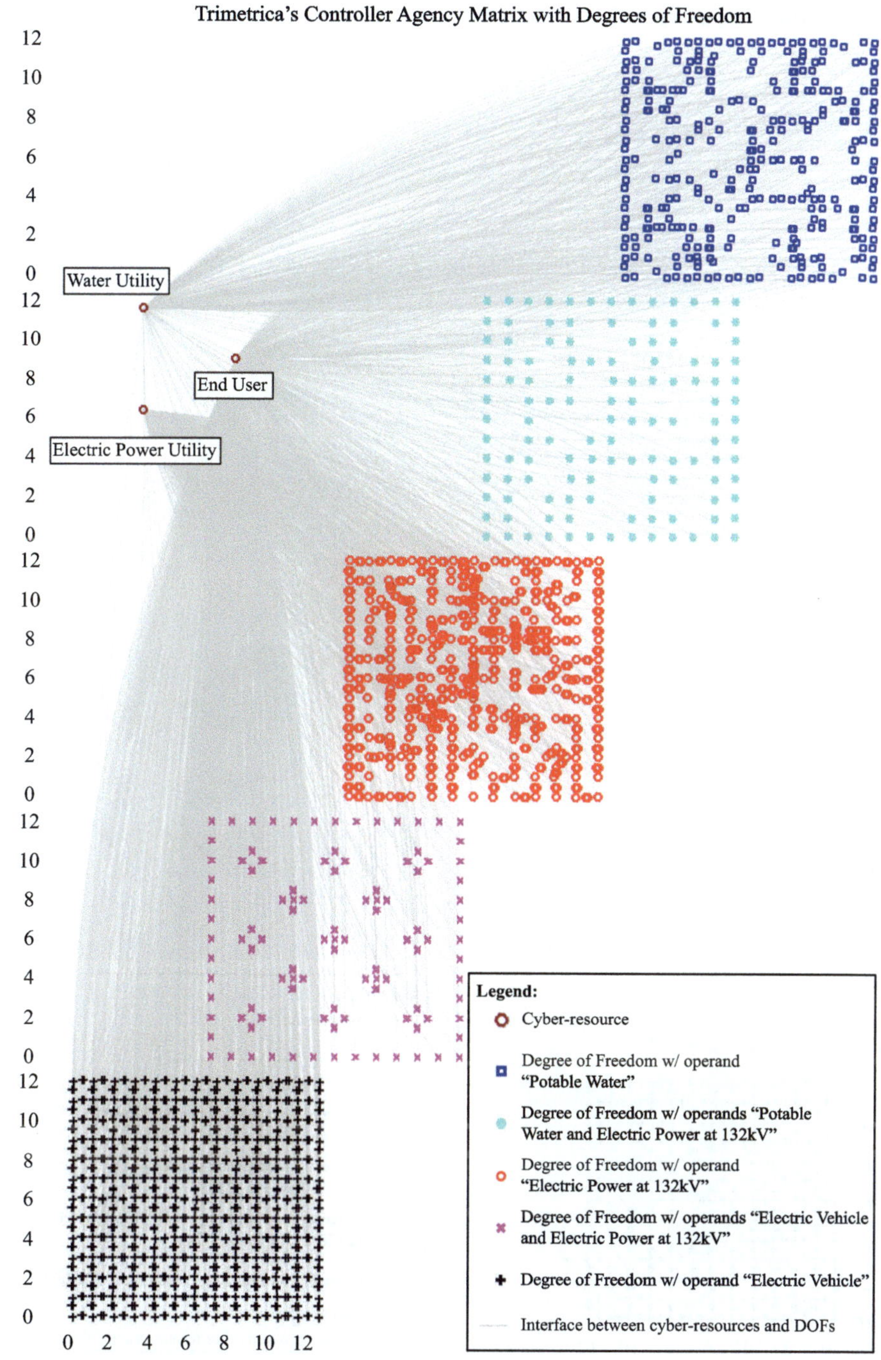

Fig. 5.24 Trimetrica's controller agency matrix: it presents the control relations between the independent cyber-resources in the top-left and the degrees of freedom under their jurisdiction

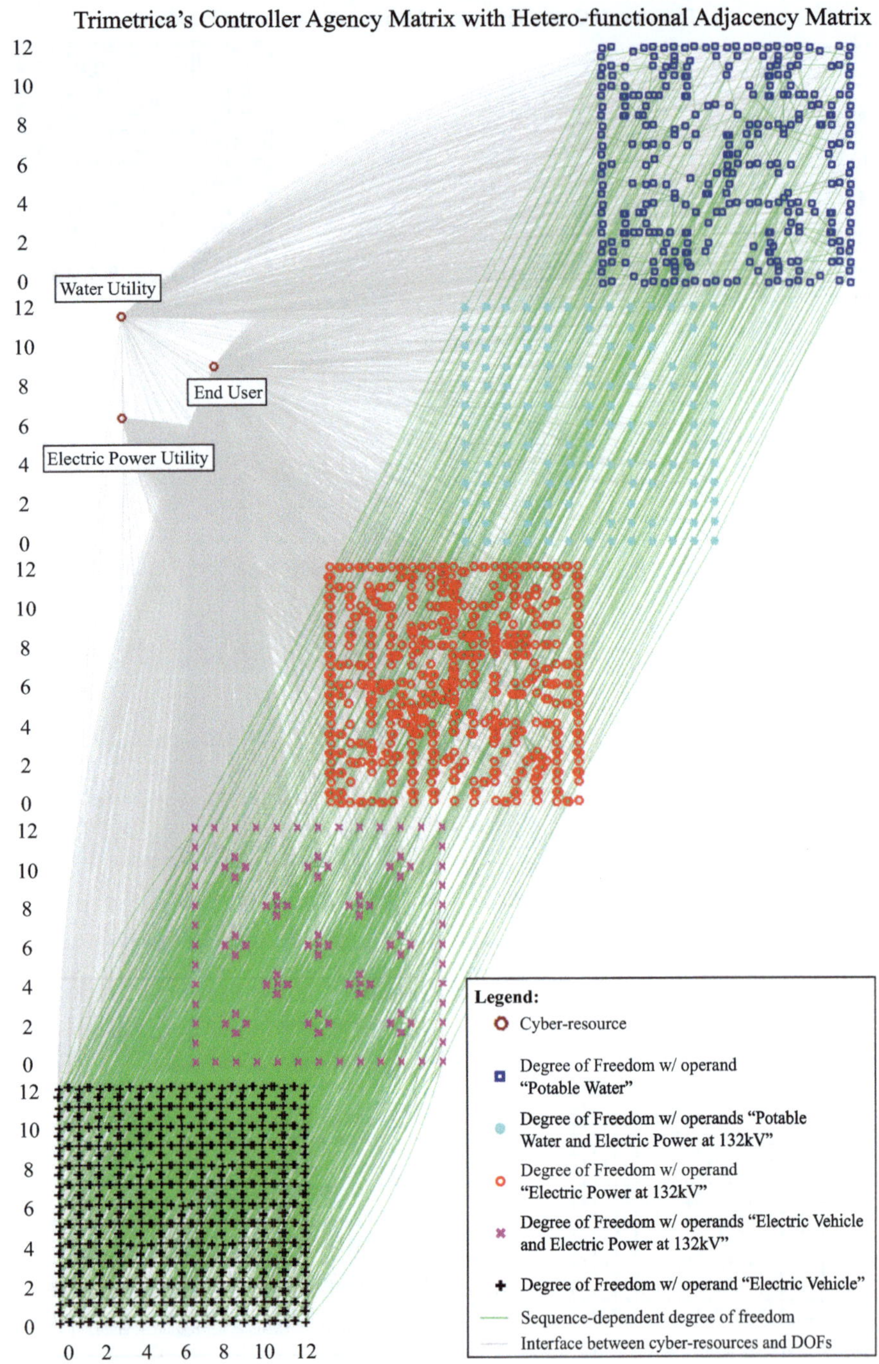

Fig. 5.25 Trimetrica's controller agency matrix superimposed on the hetero-functional adjacency matrix: the gray edges represent the control relations between the independent cyber-resources and the degrees of freedom under their jurisdiction. The green edges represent the sequence-dependent degrees of freedom as calculated in Section 5.4

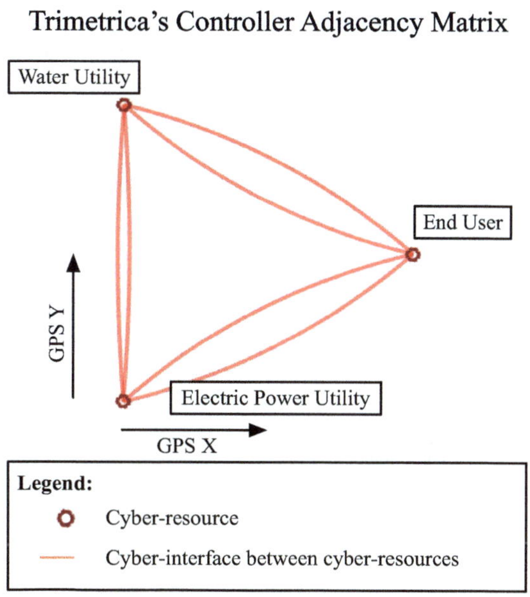

Fig. 5.26 Trimetrica's controller adjacency matrix: it presents the informatic interfaces between the cyber-resources

5.7 Service as Operand Behavior

The previous sections discussed the control structure of Trimetrica's independent smart city infrastructure system. This section continues by describing the operands in Trimetrica as they move through the city's structure to deliver services. This section draws on the theory presented in Chapter 4.5 on Page 68. The section first discusses Trimetrica's service delivery in SysML. Thereafter, it derives the service nets for Trimetrica, which are consequently translated to a service graph. The following section couples the service graph to the hetero-functional adjacency matrix.

5.7.1 Service Delivery in SysML

Trimetrica contains infrastructure to deliver three types of services in the smart city:

1. potable water for consumption of any kind.
2. electric power for consumption or work of any kind.
3. electric transportation.

As introduced in Section 4.5, the operand state can be represented in SysML using State Machines.

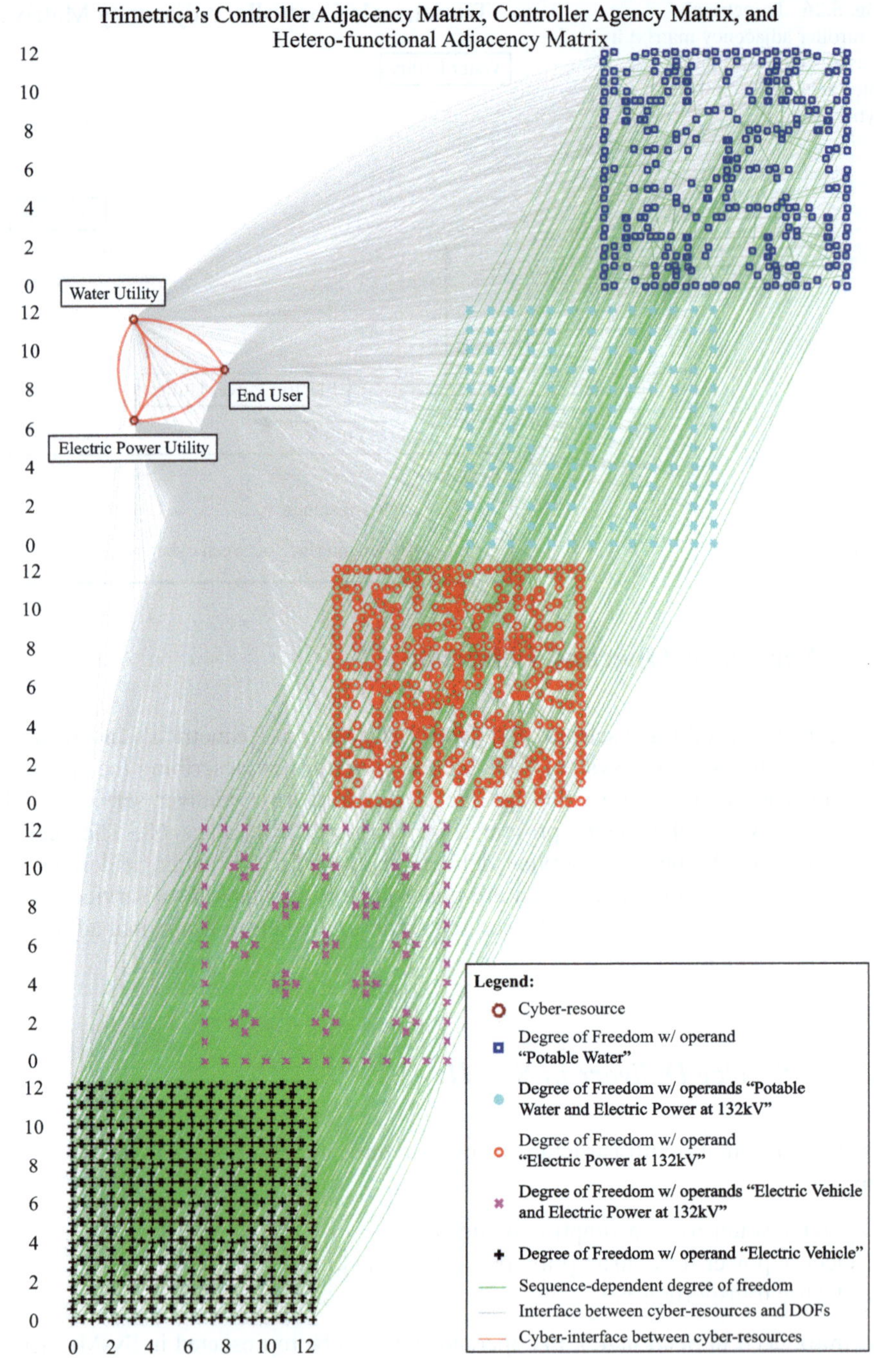

Fig. 5.27 Trimetrica's controller adjacency matrix superimposed on Figure 5.25, which includes the controller agency matrix and the hetero-functional adjacency matrix. The red edges represent the cyber-interfaces between the cyber-resources

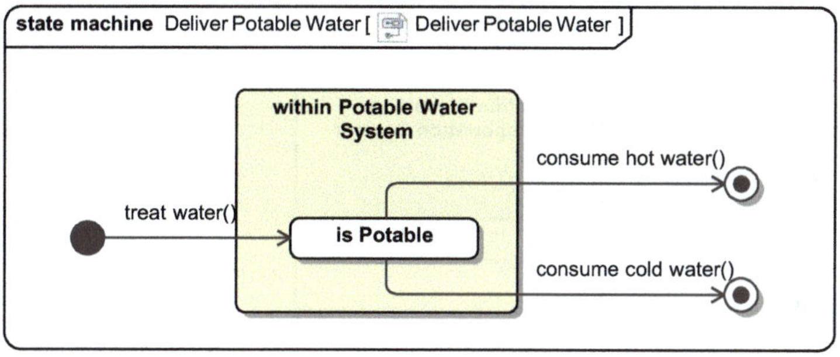

Fig. 5.28 State machine for the service "Deliver Potable Water" in the Trimetrica interdependent smart city infrastructure system

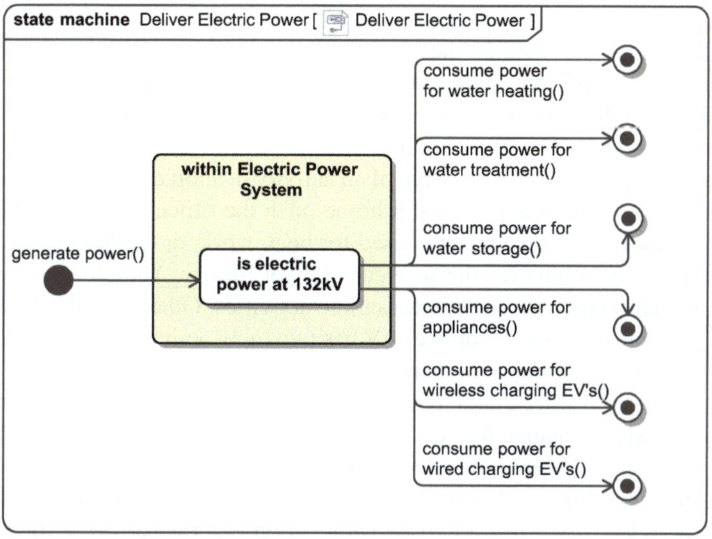

Fig. 5.29 State machine for the service "Deliver Electric Power" in the Trimetrica interdependent smart city infrastructure system

The state machine for the first service, "deliver potable water," is displayed in Figure 5.28. The state machine contains three activities. The activity *"treat water()"* achieves two goals. First, it injects water into the potable water system, changing the state of water from "outside the potable water system" to "within the potable water system." Second, the water is transformed from its input state of surface water, to the state "is potable." The activities *"consume hot water()"* and *"consume cold water()"* oppose the first activity. They withdraw potable water from the potable water system and change its state from "potable water" to "hot waste water" or "cold waste water," respectively.

The state machine for the second service, "deliver electric power," is displayed in Figure 5.29. The state machine contains seven activities. The activity *"generate*

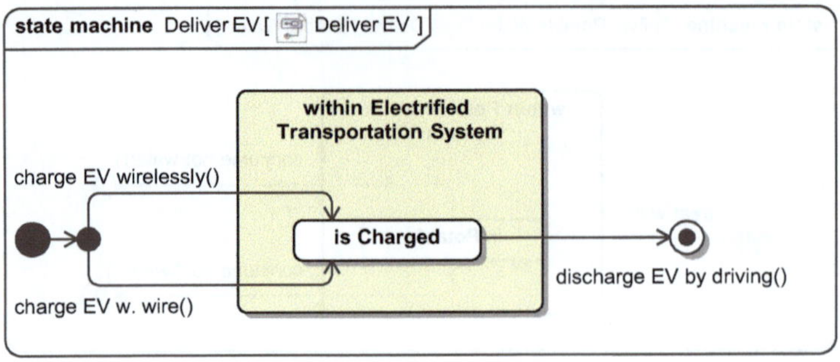

Fig. 5.30 State machine for service "Deliver Electric Vehicle" in the Trimetrica interdependent smart city infrastructure system

power()" changes the state of the energy from outside the system to "within the electric power system" and changes the state also to "electric power at 132 kV." Note that power generation would, for example, take natural gas as its input from outside the system boundary. The other six activities all withdraw electric power from the electric power system, either by virtue of an activity in another infrastructure system, or, for example, to drive appliances at home or in the office. Naturally, the state of the electric power is changed as it is used for heat, work, or to charge an EV.

The last state machine for the service "Deliver Electric Vehicle" is displayed in Figure 5.30. The state machine contains three activities. Other than the previous two services, the state machine for deliver EV has two input activities. The EV is charged either wirelessly via induction charging in the electrified roads or by wire using the charging stations throughout Trimetrica. These activities take the electric power from power grid and change the state of the EV's battery. Note that it would be more appropriate to use a continuous state for the EV because the state of its battery is not discrete. However, for simplicity of presentation the state machine representation is used. The activity *"discharge EV by driving()"* depletes the battery as the EV drives through the Trimetrica system. The work done by the vehicle is dissipated as heat and movement outside the system boundary.

5.7.2 *Service Delivery Using Petri Nets*

The SysML diagrams provide detailed insight into the nature of each of the services. However, the SysML diagrams don't lend themselves to quantitative analysis. Therefore, hetero-functional graph theory introduces Petri net based service nets. Service nets facilitate the capture of the potentially complex behavior of operands quantitatively.

The service net for the service "Deliver Potable Water" is presented in Figure 5.31. The figure shows four states of the operand with four transitions that evolve the state. The states of water are:

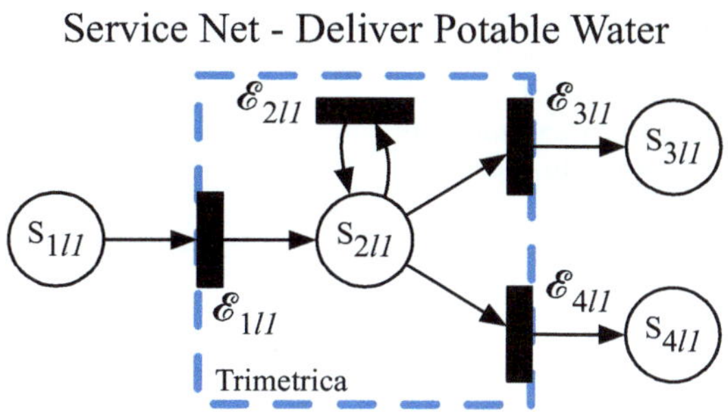

Fig. 5.31 Service net for the service "Deliver Potable Water" in the Trimetrica interdependent smart city infrastructure system

1. S_{1l_1}: Surface water
2. S_{2l_1}: Potable water
3. S_{3l_1}: Hot waste water
4. S_{4l_1}: Cold waste water

The four transitions are:

1. $\mathcal{E}_{1l_1}$: Treat water
2. $\mathcal{E}_{2l_1}$: Maintain potable water[9]
3. $\mathcal{E}_{3l_1}$: Consume hot water
4. $\mathcal{E}_{4l_1}$: Consume cold water

Note that the system boundary is indicated using a blue striped box. Transitions $\mathcal{E}_{1l_1}$, $\mathcal{E}_{3l_1}$, and $\mathcal{E}_{4l_1}$ are located on the edge of the system. They import and export the operands across the system boundary. Additionally, states S_{1l_1}, S_{3l_1}, and S_{4l_1} are depicted outside the system boundary. These states provide context to the service net and support the reader's intuition.

The service net for the service "Deliver Electric Power" is presented in Figure 5.32. The figure shows eight potential states of the operand with eight transitions that evolve the state. The states of power are:

1. S_{1l_2}: Fuel
2. S_{2l_2}: Electric power at 132 kV
3. S_{3l_2}: Work for water treatment
4. S_{4l_2}: Heat for hot water consumption
5. S_{5l_2}: Work for water storage
6. S_{6l_2}: Electric power for appliances
7. S_{7l_2}: Power for wireless charging of EVs
8. S_{8l_2}: Power for wired charging of EVs

[9]The addition of a single "maintain operand state" transition for each place is absolutely necessary once holding processes of a transformative nature are added to the model. This has been discussed on Page 71.

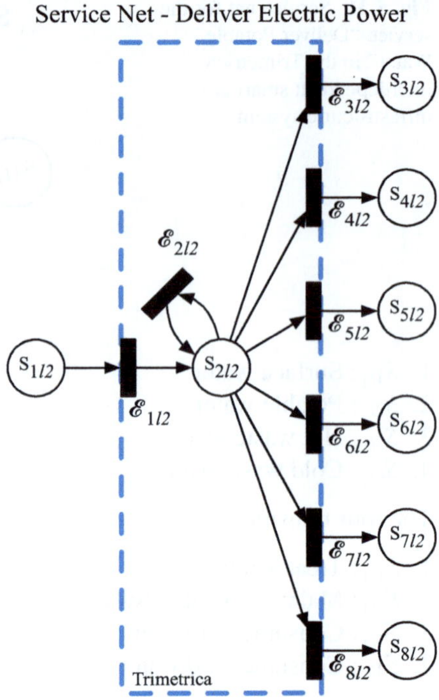

Fig. 5.32 Service net for the service "Deliver Electric Power" in the Trimetrica interdependent smart city infrastructure system

The eight transitions are:

1. $\mathcal{E}_{1l_2}$: Generate power
2. $\mathcal{E}_{2l_2}$: Maintain electric power at 132 kV
3. $\mathcal{E}_{3l_2}$: Consume power for water heating
4. $\mathcal{E}_{4l_2}$: Consume power for water treatment
5. $\mathcal{E}_{5l_2}$: Consume power for water storage
6. $\mathcal{E}_{6l_2}$: Consume power for appliances
7. $\mathcal{E}_{7l_2}$: Consume power for wireless charging of EVs
8. $\mathcal{E}_{8l_2}$: Consume power for wired charging of EVs

Note that the system boundary is indicated using a blue striped box. All transitions except $\mathcal{E}_{2l_2}$ are located on the edge of the system such that they import and export the operands across the system boundary. Additionally, all states except S_{2l_2} are depicted outside the system boundary. These states provide context to the service net and support the reader's intuition.

The final service net is for the service "Deliver EV" as presented in Figure 5.33. The figure shows four states and four transitions. The states represent the "state-of-charge" of the electric vehicle. Consequently, the states are the following:

1. S_{1l_3}: Electric power at wireless charger
2. S_{2l_3}: Electric power at wired charger
3. S_{3l_3}: State-of-charge electric vehicle
4. S_{4l_3}: Work as movement of electric vehicle

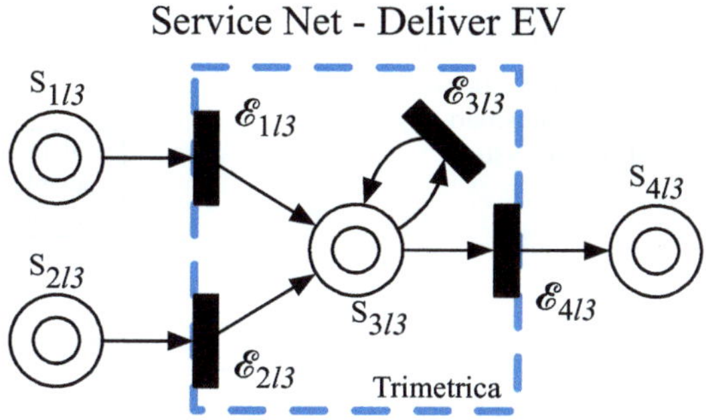

Fig. 5.33 Service net for service "Deliver Electric Vehicle" in the Trimetrica interdependent smart city infrastructure system

And the transitions are the following:

1. $\mathcal{E}_{1l_3}$: Charge EV wirelessly
2. $\mathcal{E}_{2l_3}$: Charge EV by wire
3. $\mathcal{E}_{3l_3}$: Maintain state-of-charge EV
4. $\mathcal{E}_{4l_3}$: Discharge EV by driving

Note that the states in this service net are presented differently from those presented in the previous service nets. The state S_{3l_3} represents state-of-charge in the battery of an electric vehicle as a rational number rather than as an integer. Consequently, a different type of Petri net is necessary [5]. Continuous Petri nets allow for the division of markers into tokens to emulate a continuous behavior. It consequently facilitates the non-discrete behavior of the electric vehicle's state of charge. The remaining states in the system are outside the system boundary and are also considered continuous for consistency.

5.7.3 Service Delivery as Service Graph

Thus far, services have been modeled as service (Petri) nets. Service activities have been represented as transitions. As discussed in Section 4.5.2, the adjacency of these transitions is calculated using a dual adjacency matrix:

$$A_{Tl_i} = M_{l_i}^{+T} M_{l_i}^{-} \tag{5.15}$$

where M^+ is the positive incidence matrix of the service net, M^- is the negative incidence matrix of the service net, and both have a size of $\sigma(S_{l_i}) \times \sigma(\mathcal{E}_{l_i})$. The resulting matrix A_{Tl_i} has a size of $\sigma(\mathcal{E}_{l_i}) \times \sigma(\mathcal{E}_{l_i})$, and shows the feasible sequences in the service net. Visually, such a graph has the transitions as nodes and the directed arcs represent their adjacency. Each of the transition adjacency matrices is now calculated. Note that the incidence matrices only consider the places and transitions *within* the system boundary.

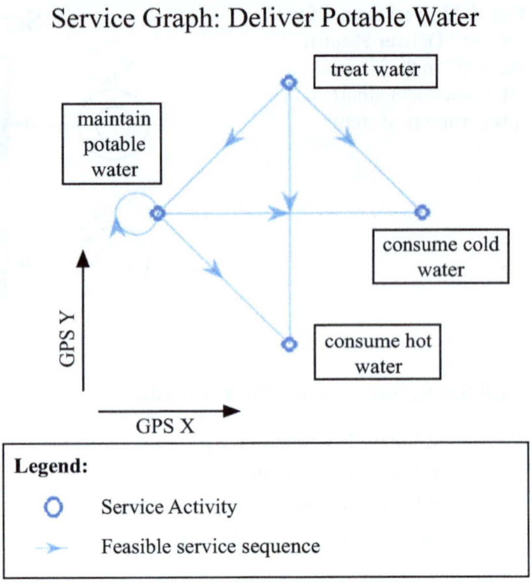

Fig. 5.34 Service graph for the service "Deliver Potable Water" in the Trimetrica interdependent smart city infrastructure system

The service "Deliver potable water" has an incidence matrix of size $\sigma(S_{l_1}) \times \sigma(\mathcal{E}_{l_1}) = 1 \times 4$. The size of the transition adjacency matrix A_{Tl_1} is $\sigma(\mathcal{E}_{l_1}) \times \sigma(\mathcal{E}_{l_1}) = 4 \times 4$ with six filled elements. Figure 5.34 shows the visualization of A_{Tl_1}. When compared with the service net in Figure 5.31, the service graph closely resembles its transition structure.

The service "Deliver electric power" has an incidence matrix of size $\sigma(S_{l_2}) \times \sigma(\mathcal{E}_{l_2}) = 1 \times 8$. The size of the transition adjacency matrix A_{Tl_2} is $\sigma(\mathcal{E}_{l_2}) \times \sigma(\mathcal{E}_{l_2}) = 8 \times 8$ with 14 filled elements. Figure 5.35 shows the visualization of A_{Tl_2}. When compared with the service net in Figure 5.32, the service graph closely resembles its transition structure.

The service "Deliver electric vehicle" has an incidence matrix of size $\sigma(S_{l_3}) \times \sigma(\mathcal{E}_{l_3}) = 1 \times 4$. The size of the transition adjacency matrix A_{Tl_3} is $\sigma(\mathcal{E}_{l_3}) \times \sigma(\mathcal{E}_{l_3}) = 4 \times 4$ with six filled elements. Figure 5.36 (on Page 147) shows the visualization of A_{Tl_3}. When compared with the service net in Figure 5.33, the service graph closely resembles its transition structure.

5.8 Service Feasibility Matrix

This section defines Trimetrica's service feasibility matrix as the coupling of the system services to the system transformation processes. Section 4.6 on Page 73 first introduced the service feasibility matrix. This section first provides the SysML description of the coupling between system structure and service delivery. Thereafter, it uses hetero-functional graph theory to describe the coupling explicitly.

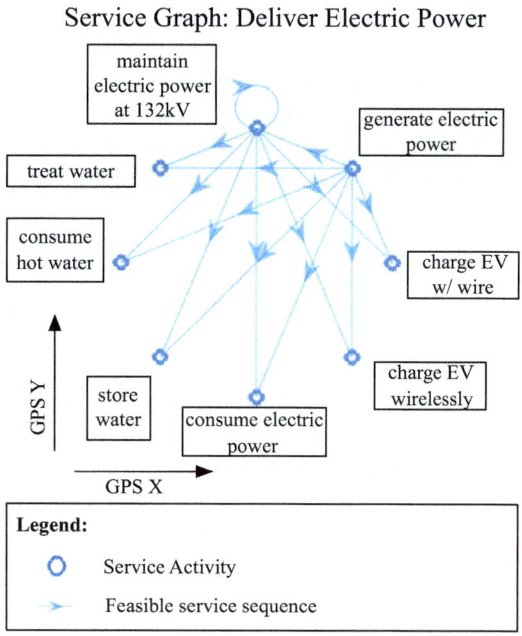

Fig. 5.35 Service graph for the service "Deliver Electric Power" in the Trimetrica interdependent smart city infrastructure system

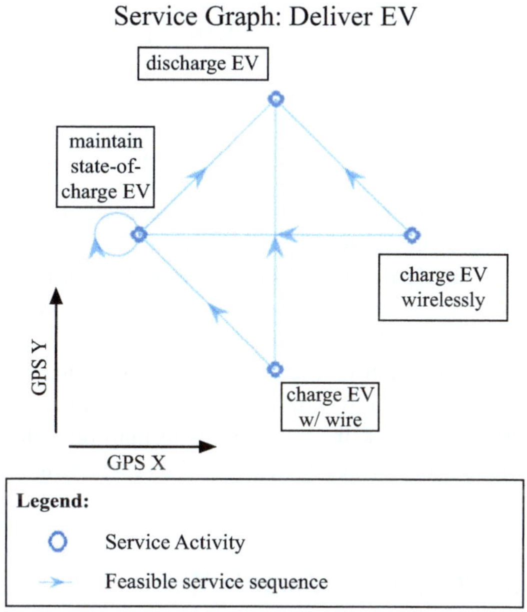

Fig. 5.36 Service graph for the service "Deliver Electric Vehicle" in the Trimetrica interdependent smart city infrastructure system

SysML generally uses state machines to describe the state of operands in the system, as discussed in Section 5.7. The activities of the system are described in the activity diagram. However, the state machines are not *explicitly* coupled to the activity diagram. The coupling of the activity diagram and the state machine is a result of both models describing the same system from different viewpoints. A triggered transformation process in the activity diagram evolves the state of the operand in the state machine. The coupling in SysML is, therefore, implicit.

Hetero-functional Graph Theory, however, describes the coupling between the service transitions and the transformation processes explicitly using the service feasibility matrix. Section 5.7 defined a separate service net for each service in Trimetrica. Consequently, each of these services has a separate service feasibility matrix as well. The size of the service feasibility matrix is defined in Section 4.6 in Definition 4.25 on Page 74 as $\sigma(\mathcal{E}_{l_i}) \times \sigma(P)$ where $l_i \in L$ is the selected service. Its elements equal 1 if e_{xl_i} realizes system process p_w.

Thus far, the service feasibility matrix has been defined such that it couples the service activities to the system processes. However, the system structure has been defined using degrees of freedom as a mapping of system function onto system form. Consequently, by mapping the service feasibility matrix onto the system resources, the coupling allows for a direct link between the capabilities in the system and the service activities, as demonstrated in Equation 5.17. The resulting matrix $\widehat{\Lambda}_i$ has size $\sigma(\mathcal{E}_{l_i}) \times \sigma(R)\sigma(P)$. For the purpose of this work, this matrix can be projected to map the service activities to the projected set of capabilities, as follows in Equation 5.18. Matrix $\widetilde{\Lambda}_i$ has size $\sigma(\mathcal{E}_{l_i}) \times \sigma(\mathcal{E}_S)$.

$$\Lambda_{iSC} = \left[\Lambda_{\mu i} \mid \Lambda_{\gamma i} \otimes \mathbb{1}^{\sigma(P_\eta)T}\right] \tag{5.16}$$

$$\widehat{\Lambda}_{iSC} = \mathbb{1}^{\sigma(R)T} \otimes \Lambda_{iSC} \tag{5.17}$$

$$\widetilde{\Lambda}_{iSC} = \left(\mathbb{1}^{\sigma(R)T} \otimes \Lambda_{iSC}\right) \mathbb{P}_S^T \tag{5.18}$$

The section now continues to define each of the projected service feasibility matrices for the three services in Trimetrica.

The derivation of the service feasibility matrices follows four steps: first, the service transformation feasibility matrix and service transportation feasibility matrix are derived for each of the services. Second, Equation 5.16 is used to calculate each of the service feasibility matrices. Third, Equation 5.17 maps the service feasibility matrix onto the system adjacency matrix. The last step projects the service feasibility matrices to map onto the degrees of freedom. Each of these steps is performed for each of the services in Trimetrica.

5.8.1 Deliver Potable Water

The service transformation and transportation feasibility matrices for the service "deliver potable water" are derived from the mapping of service activities onto system processes. The set of service transitions for the service "deliver potable water" has a size of four and is defined as: $\mathcal{E}_{l_1} = \{$treat water, maintain potable water, consume hot water, consume cold water}. The set of transformation processes in Trimetrica is $P_{\mu SC}$ = {treat water, generate electric power, consume hot water, consume cold water, consume electric power}. The service transformation feasibility matrix $\Lambda_{\mu l_1}$ has a size of $\sigma(\mathcal{E}_{l_1}) \times P_{\mu SC} = 4 \times 5$ and contains three filled elements. The set of holding processes in Trimetrica is $P_{\gamma SC}$ = {carry potable water, carry potable water while consuming electric power, carry electric power at 132 kV, carry electric vehicle without affecting battery, carry electric vehicle while discharging, carry electric vehicle while charging by wire, carry electric vehicle while charging wirelessly}. The service transportation feasibility matrix $\Lambda_{\gamma l_1}$ has a size of $\sigma(\mathcal{E}_{l_1}) \times P_{\gamma SC} = 4 \times 7$ and contains one filled element.

Based on the service transformation and transportation feasibility matrix, Equations 5.16, 5.17, and 5.18 calculate the (1) service feasibility matrix, (2) service feasibility matrix projected to the system adjacency matrix, and (3) service feasibility matrix projected to the degrees of freedom, respectively:

1. Λ_{l_1} has a size of $\sigma(\mathcal{E}_{l_1}) \times \sigma(P_{SC}) = 4 \times 703{,}428$. It contains 100,492 filled elements.
2. $\widehat{\Lambda}_{1SC}$ has a size of $\sigma(\mathcal{E}_{l_1}) \times \sigma(R_{SC})\sigma(P_{SC}) = 4 \times 860{,}292{,}444$. It contains 122,901,716 filled elements.
3. $\widetilde{\Lambda}_{1SC}$ has a size of $\sigma(\mathcal{E}_{l_1}) \times \sigma(\mathcal{E}_S) = 4 \times 1991$. It contains 430 filled elements.

5.8.2 Deliver Electric Power

The service transformation and transportation feasibility matrices for the service "deliver electric power" are derived from the mapping of service activities onto system processes. The set of service transitions has a size of eight and is defined as: $\mathcal{E}_{l_2}$ = {generate electric power, maintain potable water, treat water, consume hot water, store water, consume electric power, charge EV wirelessly, charge EV w/ wire}. The set of transformation processes in Trimetrica is $P_{\mu SC}$ = {treat water, generate electric power, consume hot water, consume cold water, consume electric power}. The service transformation feasibility matrix $\Lambda_{\mu l_2}$ has a size of $\sigma(\mathcal{E}_{l_2}) \times P_{\mu SC} = 8 \times 5$ and contains four filled elements. The set of holding processes in Trimetrica is $P_{\gamma SC}$ = {carry potable water, carry potable water while consuming electric power, carry electric power at 132 kV, transport electric vehicle without affecting battery, transport electric vehicle while discharging, carry electric vehicle while charging by wire, carry electric vehicle while charging wirelessly}. The service transportation feasibility matrix $\Lambda_{\gamma l_2}$ has a size of $\sigma(\mathcal{E}_{l_2}) \times P_{\gamma SC} = 8 \times 7$ and contains four filled elements.

Based on the service transformation and transportation feasibility matrix, Equations 5.16, 5.17, and 5.18 calculate the (1) service feasibility matrix, (2) service feasibility matrix projected to the system adjacency matrix, and (3) service feasibility matrix projected to the degrees of freedom, respectively:

1. Λ_{l_2} has a size of $\sigma(\mathcal{E}_{l_2}) \times \sigma(P_{SC}) = 8 \times 703{,}428$. It contains 401,960 filled elements.
2. $\widehat{\Lambda}_{2SC}$ has a size of $\sigma(\mathcal{E}_{l_2}) \times \sigma(R_{SC})\sigma(P_{SC}) = 8 \times 860{,}292{,}444$. It contains 491,597,080 filled elements.
3. $\widetilde{\Lambda}_{2SC}$ has a size of $\sigma(\mathcal{E}_{l_2}) \times \sigma(\mathcal{E}_S) = 8 \times 1991$. It contains 883 filled elements.

5.8.3 Deliver Electric Vehicle

The service transformation and transportation feasibility matrices for the service "deliver electric vehicle" are derived from the mapping of service activities onto system processes. The set of service transitions has a size of four and is defined as: $\mathcal{E}_{l_3} = \{$charge EV wirelessly, charge EV w/ wire, maintain state-of-charge EV, discharge EV$\}$. The set of transformation processes in Trimetrica is $P_{\mu SC} = \{$treat water, generate electric power, consume hot water, consume cold water, consume electric power$\}$. The service transformation feasibility matrix $\Lambda_{\mu l_3}$ has a size of $\sigma(\mathcal{E}_{l_3}) \times P_{\mu SC} = 4 \times 5$ and contains zero filled elements. The set of holding processes in Trimetrica is $P_{\gamma SC} = \{$carry potable water, carry potable water while consuming electric power, carry electric power at 132 kV, transport electric vehicle without affecting battery, transport electric vehicle while discharging, carry electric vehicle while charging by wire, carry electric vehicle while charging wirelessly$\}$. The service transportation feasibility matrix $\Lambda_{\gamma l_3}$ has a size of $\sigma(\mathcal{E}_{l_3}) \times P_{\gamma SC} = 4 \times 7$ and contains four filled elements.

Based on the service transformation and transportation feasibility matrix, Equations 5.16, 5.17, and 5.18 calculate the (1) service feasibility matrix, (2) service feasibility matrix projected to the system adjacency matrix, and (3) service feasibility matrix projected to the degrees of freedom, respectively:

1. Λ_{l_3} has a size of $\sigma(\mathcal{E}_{l_3}) \times \sigma(P_{SC}) = 4 \times 703{,}428$. It contains 401,956 filled elements.
2. $\widehat{\Lambda}_{3SC}$ has a size of $\sigma(\mathcal{E}_{l_3}) \times \sigma(R_{SC})\sigma(P_{SC}) = 4 \times 860{,}292{,}444$. It contains 491,592,188 filled elements.
3. $\widetilde{\Lambda}_{3SC}$ has a size of $\sigma(\mathcal{E}_{l_3}) \times \sigma(\mathcal{E}_S) = 4 \times 1991$. It contains 950 filled elements.

5.8.4 Visualizing the Service Feasibility Matrix

As the service feasibility matrices describe the interface between the service graphs and the smart city capabilities, they can be represented as bipartite graphs. The rows of each of the service feasibility matrices represent the service transitions, and the columns represent the degrees of freedom. Since the service feasibility matrix is a bipartite graph, it can be visualized. Figure 5.37 on Page 151 couples the service graphs to the system capabilities, in a way similar to the coupling of the controller

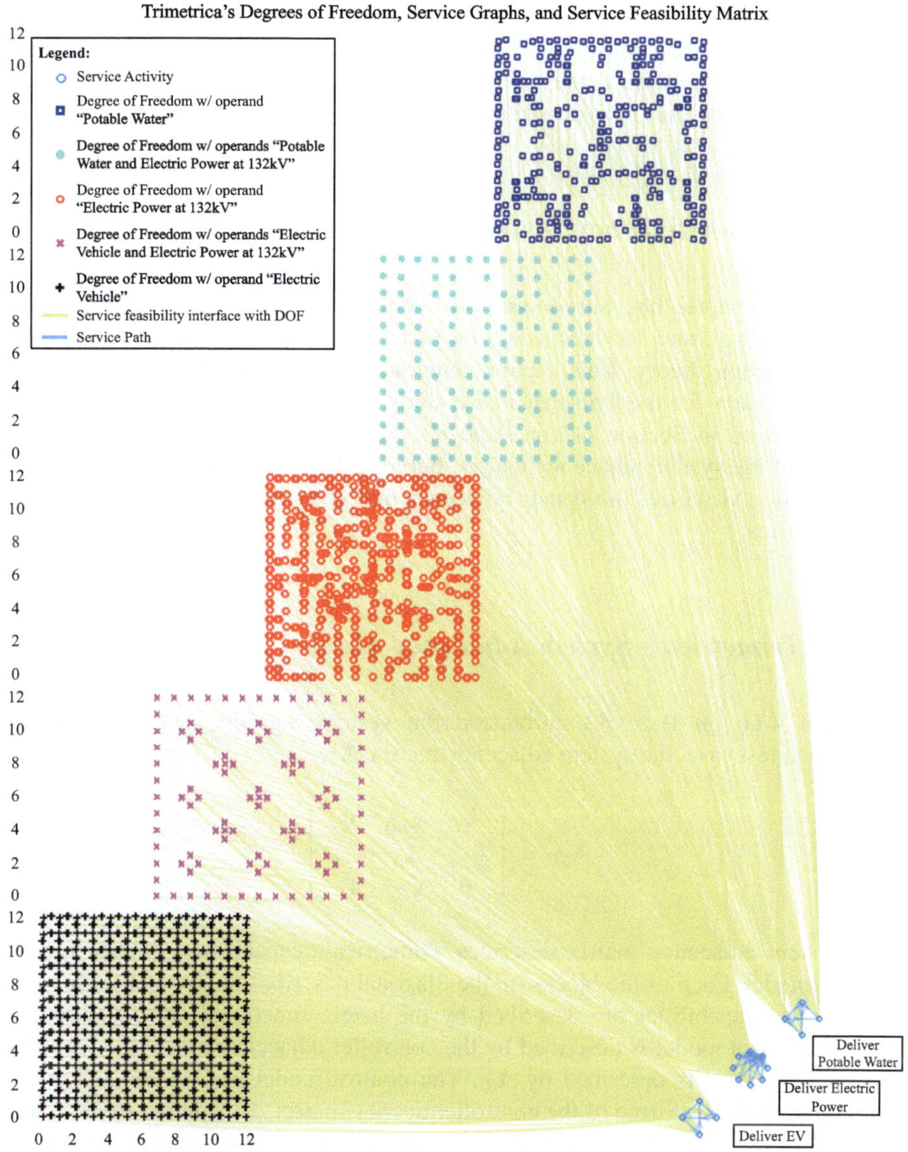

Fig. 5.37 Trimetrica's service graphs and the service feasibility matrix: the service graphs are represented in the bottom-right corner of the figure and are drawn from Figures 5.34, 5.35, and 5.36. The service feasibility matrix is the interface shown in yellow between the service graphs and the degrees of freedom

agents to the system capabilities. Note that many of the degrees of freedom realize more than one service transition. For example, the capability "treat water at water treatment facility 1" realizes the service transition "treat water" in both the "deliver potable water" and the "deliver electric power" services.

5.9 System Adjacency Matrix

So far, this chapter has introduced the Trimetrica test case, and described its structure, control, and services using the first six mathematical models in hetero-functional graph theory. This section integrates the six models into the system adjacency matrix for the Trimetrica case study. The system adjacency matrix was first introduced in Section 4.7 on Page 82. This section first briefly revisits the elements of the system adjacency matrix, before going into detail. The second part of this section visualizes the system adjacency matrix as the hetero-functional graph for Trimetrica.

5.9.1 *Trimetrica's System Adjacency Matrix*

Equation 4.66 on Page 82 introduced the system adjacency matrix. For the Trimetrica test case, the system adjacency matrix $\mathbb{A}_{SC}$ is:

$$\mathbb{A}_{SC} = \begin{bmatrix} \mathbb{A}_L & \mathbb{A}_{L\rho} & \mathbf{0} \\ \mathbb{A}_{\rho L} & A_\rho & \mathbb{A}_{\rho C} \\ \mathbf{0} & \mathbb{A}_{C\rho} & A_C \end{bmatrix} \tag{5.19}$$

The system adjacency matrix describes Trimetrica's capabilities, control model, service model. Each of the blocks on the diagonal describes the core of those three pillars. The capabilities are described by the hetero-functional adjacency matrix A_ρ, the control model is described by the controller adjacency matrix A_C, and the operand behavior is described by $\mathbb{A}_L$. The control model is then coupled to the structural model by virtue of the controller agency matrix, as captured by $\mathbb{A}_{\rho C}$ and $\mathbb{A}_{C\rho}$. The service model is coupled to the structural model by virtue of the service feasibility matrix $\mathbb{A}_{\rho L}$ and its transpose. Equation 5.20 calculates the projected system adjacency matrix for the Trimetrica test case. This section continues to use the projected system adjacency matrix for its calculations.

$$\widetilde{\mathbb{A}}_{SC} = \begin{bmatrix} \mathbb{A}_L & \widetilde{\mathbb{A}}_{L\rho} & \mathbf{0} \\ \widetilde{\mathbb{A}}_{\rho L} & \widetilde{A}_{\rho SC} & \widetilde{\mathbb{A}}_{\rho C} \\ \mathbf{0} & \widetilde{\mathbb{A}}_{C\rho} & A_{CSC} \end{bmatrix} \tag{5.20}$$

Block $\widetilde{A}_{\rho SC}$: The hetero-functional adjacency matrix is the core of the system adjacency matrix, because it contains all system capabilities and their feasible sequences. It is defined in Section 5.4 on page 125. As mentioned previously, because the system adjacency matrix is projected, it dictates the sizes of the coupling matrices. The size of $\widetilde{A}_\rho$ is $\mathcal{E}_S \times \mathcal{E}_S = 1991 \times 1991$, and the number of filled elements in $\widetilde{A}_\rho$ is the number of sequence dependent constraints: 8082.

Block A_{CSC}: The controller adjacency matrix describes the cyber-interfaces between the cyber-resources in Trimetrica. It is calculated in Section 5.6 on page 136. The controller adjacency matrix has size $\sigma(Q_{SC}) \times \sigma(Q_{SC}) = 3 \times 3$. The number of filled elements is 9.

Block $\mathbb{A}_L$: The upper left block of the system adjacency matrix describes the service model. It takes the service activities as nodes and shows the logical coupling between them. Equation 4.68 on Page 84 introduces the block diagonal form of the matrix so that each service behavior is uncoupled for the next. For the Trimetrica test case, matrix $\mathbb{A}_L$ is:

$$\mathbb{A}_L = \begin{bmatrix} A_{Tl_1} & 0 & 0 \\ 0 & A_{Tl_2} & 0 \\ 0 & 0 & A_{Tl_3} \end{bmatrix} \tag{5.21}$$

The elements A_{Tl_i} are calculated in Section 5.7.3 on Page 145. The size of $\mathbb{A}_L$ is $\sum_{i=1}^{3} \sigma(\mathcal{E}_{l_i}) \times \sum_{i=1}^{3} \sigma(\mathcal{E}_{l_i}) = 16 \times 16$. The number of filled elements is 26.

Block $\widetilde{\mathbb{A}}_{\rho C}$ **and** $\widetilde{\mathbb{A}}_{C\rho}$: These two block matrices couple the structural degrees of freedom in $A_{\rho SC}$ and the independent cyber-resources Q_{SC}. Equation 5.14 on Page 135 calculates the projected controller agency matrix for the Trimetrica test case:

$$\widetilde{\mathbb{A}}_{\rho C} = \widetilde{A}_{QSC} = \mathbb{P}(\overline{A}_{QSC} \otimes \mathbb{1}^{\sigma(P)}) \tag{5.22}$$

where $\widetilde{A}_{\rho C}$ has a size of $\mathcal{E}_S \times \sigma(Q_{SC}) = 1991 \times 3$. The matrix contains 1991 filled elements. The Trimetrica test case does not differentiate between sensing and actuation signals, and, therefore, matrix $\widetilde{\mathbb{A}}_{C\rho}$ is:

$$\widetilde{\mathbb{A}}_{C\rho} = \widetilde{\mathbb{A}}_{\rho C}^{T} \tag{5.23}$$

where $\widetilde{\mathbb{A}}_{C\rho}$ has size $\sigma(Q_{SC}) \times \mathcal{E}_S = 3 \times 1991$ with 1991 filled elements.

Block $\mathbb{A}_{L\rho}$ **and** $\mathbb{A}_{\rho L}$: These matrices capture the coupling of Trimetrica's service nets with the system structure. In Section 5.8, the service feasibility matrices were separately introduced for each service. However, the system adjacency matrix includes the service nets as a combination of each of the separate service graphs, which requires that the service feasibility matrices are combined as well. Using

Equation 4.72 on Page 84, the concatenation of the service feasibility matrices follows:

$$\mathbb{A}_{L\rho} = (\mathbb{A}_{\rho L})^T = \begin{bmatrix} \widehat{\Lambda}_{1SC} \\ \widehat{\Lambda}_{2SC} \\ \widehat{\Lambda}_{3SC} \end{bmatrix} \tag{5.24}$$

$$\widetilde{\mathbb{A}}_{L\rho} = (\widetilde{\mathbb{A}}_{\rho L})^T = \begin{bmatrix} \widetilde{\Lambda}_{1SC} \\ \widetilde{\Lambda}_{2SC} \\ \widetilde{\Lambda}_{3SC} \end{bmatrix} \tag{5.25}$$

where $\widetilde{\mathbb{A}}_{L\rho}$ has a size of $\sum_{i=1}^{3} \sigma(\mathcal{E}_{l_i}) \times \mathcal{E}_S = 16 \times 1991$ with $430+883+950 = 2263$ filled elements.

In conclusion, after the discussion of each of the seven block matrices within the system adjacency matrix, the total number of filled elements is calculated to be: $8082 + 9 + 26 + 1991 + 1991 + 2263 + 2263 = 16{,}625$. The size of the system adjacency matrix $\mathbb{A}_{SC}$ is:

$$\left[\sum_{i=1}^{3} \sigma(\mathcal{E}_{l_i}) + \sigma(R_P)\sigma(P) + \sigma(Q)\right] \times \left[\sum_{i=1}^{3} \sigma(\mathcal{E}_{l_i}) + \sigma(R_P)\sigma(P) + \sigma(Q)\right]$$
$$= 860{,}292{,}463 \times 860{,}292{,}463 \tag{5.26}$$

And the size of the projected system adjacency matrix $\widetilde{\mathbb{A}}_{SC}$ is:

$$\left[\sum_{i=1}^{3} \sigma(\mathcal{E}_{l_i}) + \sigma(\mathcal{E}_S) + \sigma(Q)\right] \times \left[\sum_{i=1}^{3} \sigma(\mathcal{E}_{l_i}) + \sigma(\mathcal{E}_S) + \sigma(Q)\right] = 2010 \times 2010 \tag{5.27}$$

5.9.2 Hetero-functional Graph Visualization

The system adjacency matrix describes the full Trimetrica interdependent smart city infrastructure test case in a single adjacency matrix. Consequently, the matrix can be represented as a graph. Intuitively, this chapter has presented pieces of the system adjacency matrix as each of the couplings were made. Figure 5.23 presented the hetero-functional adjacency matrix as a five-layer network, with degrees of freedom as its nodes, and sequence-dependent degrees of freedom as the edges. Thereafter, the controller agency matrix coupled the cyber-resources to the degrees of freedom graphically in Figure 5.27. After the control structure, the operand behavior was defined and coupled to the hetero-functional adjacency matrix as depicted in Figure 5.37. Finally, these different visualizations are pieced together in Figure 5.38, which presents the hetero-functional graph, a visual representation of the system adjacency matrix.

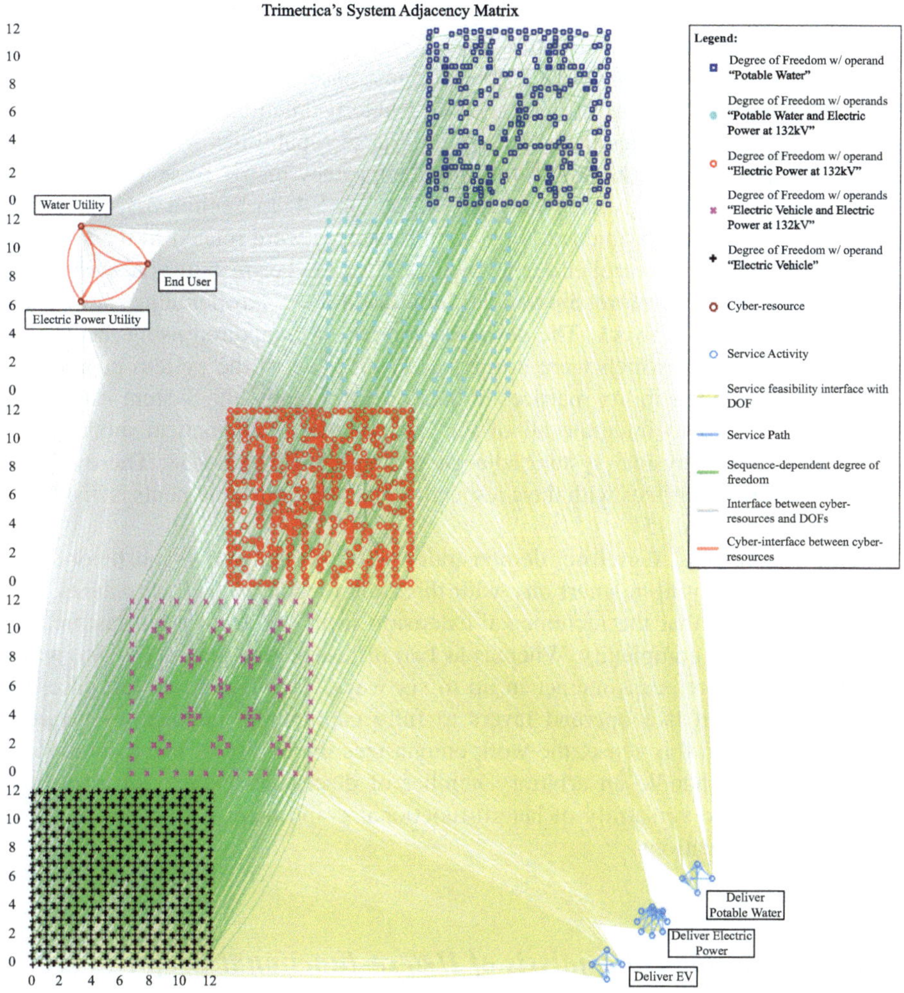

Fig. 5.38 The system adjacency matrix for the Trimetrica interdependent smart city infrastructure system presented as a hetero-functional graph

5.10 Discussion

This chapter has demonstrated the theoretical discussion of hetero-functional graph theory on an illustrative example. It has shown that an interdependent smart city infrastructure system with an arbitrary topology can be modeled using hetero-functional graph theory. First, the chapter provided a background of the development of a test case in Section 5.1. The primary advantage of test cases is that they facilitate the standardized comparison of different modeling and analysis methods. In Section 5.2, the Trimetrica test case was introduced. It delivers three services to the residents of the city by virtue of an interdependent physical infrastructure and a control structure.

After the introduction of the Trimetrica test case, Section 5.3 calculated the system knowledge base as the first model of hetero-functional graph theory so as to identify the capabilities in the smart city. These capabilities were consequently coupled sequentially in the hetero-functional adjacency matrix so as to identify the sequence-dependent degrees of freedom in Section 5.4. In Section 5.5, the cyber-resources that control Trimetrica's physical resources were related to the system's capabilities within a controller agency matrix. These cyber-resources also interface with one another. Therefore, Section 5.6 introduced the controller adjacency matrix as a type of social network. The delivered services are captured using the service nets in Section 5.7, which were consequently coupled to the system capabilities using the service feasibility matrices in Section 5.8. Finally, the System Adjacency Matrix succeeded to integrate all of the independent mathematical models, and generate a cyber-physical system adjacency matrix in Figure 5.38. The overview of the respective matrices with their respective sizes and filled elements is displayed in Table 5.2 on page 157.

This chapter has, therefore, demonstrated hetero-functional graph theory as a means of representing a smart city with three interdependent infrastructures. The work also shows that the incremental extension from two to three infrastructures is of permutation complexity. Whereas as two infrastructures interact in two ways, three infrastructures can interact in up to six ways. The Trimetrica test case, for example, required five operand layers to fully capture the interactions between infrastructures. In this sense, the work emphasizes the ability of hetero-functional graph theory to handle an arbitrary number of discipline-specific infrastructure. Finally, the broad versatility of hetero-functional graph theory is supported by its ontological foundations.

5.10.1 Ontological Analysis of Hetero-functional Graph Theory

The foundation of hetero-functional graph theory is specifically built on the ontological properties of soundness, completeness, lucidity, and laconicity, introduced in Section 3.1 on page 23. Recall that the traditional application of graph theory violates the properties of completeness and lucidity as it (1) fails to represent the complete set of concepts in the domain abstraction with modeling primitives, and (2) overloads modeling primitives with multiple domain concepts.

Hetero-functional graph theory maintains the four ontological properties to ensure an isomorphic representation of the conceptual abstraction. The diverse nature of large flexible engineering systems require a diverse language, and the seven models in hetero-functional graph theory provide the necessary breadth. All LFESs consist of a structural model, a controller model, and an operand behavior

Table 5.2 An overview of Trimetrica's seven mathematical models of hetero-functional graph theory

Model and section	Matrix	Size	Filled elements
Degrees of freedom/capabilities Section 5.3 on page 101	A_{SSC}	$\sigma(P) \times \sigma(R_P) = 703{,}428 \times 1223$	Degrees of freedom: 1991
Hetero-functional adjacency matrix Section 5.4 on page 125	$A_{\rho SC}$	$\sigma(R_P)\sigma(P) \times \sigma(R_P)\sigma(P) = 860{,}292{,}444 \times 860{,}292{,}444$	Sequence-dependent degrees of freedom: 8082
	$\widetilde{A}_{\rho SC}$	$\sigma(\mathcal{E}_S) \times \sigma(\mathcal{E}_S) = 1991 \times 1991$	Sequence-dependent degrees of freedom: 8082
Controller agency matrix Section 5.5 on page 131	A_{QSC}	$\sigma(R_P) \times \sigma(R) = 1223 \times 1226$	Cyber-physical interfaces: 1223
	$\overline{A}_{QSC}$	$\sigma(R_P) \times \sigma(Q) = 1223 \times 3$	Cyber-physical interfaces: 1223
	$\widetilde{A}_{QSC}$	$\sigma(\mathcal{E}_S) \times \sigma(Q) = 1991 \times 3$	Cyber-physical interfaces: 1991
Controller adjacency matrix Section 5.6 on page 136	A_{CSC}	$\sigma(Q) \times \sigma(Q) = 3 \times 3$	Cyber-interfaces: 9
Service as operand behavior Section 5.7 on page 139	A_{Tl_1}	$\sigma(\mathcal{E}_{l_1}) \times \sigma(\mathcal{E}_{l_1}) = 4 \times 4$	Service sequences: 6
	A_{Tl_2}	$\sigma(\mathcal{E}_{l_2}) \times \sigma(\mathcal{E}_{l_2}) = 8 \times 8$	14
	A_{Tl_3}	$\sigma(\mathcal{E}_{l_3}) \times \sigma(\mathcal{E}_{l_3}) = 4 \times 4$	6
	$\mathbb{A}_L$	$\sum_{i=1}^{3} \sigma(\mathcal{E}_{l_i}) \times \sum_{i=1}^{3} \sigma(\mathcal{E}_{l_i}) = 16 \times 16$	Service sequences: 26
Service feasibility matrix Section 5.8 on page 146	Λ_{l_i}	$\sigma(\mathcal{E}_{l_1}) \times \sigma(P) = 4 \times 703{,}428$ $\sigma(\mathcal{E}_{l_2}) \times \sigma(P) = 8 \times 703{,}428$ $\sigma(\mathcal{E}_{l_3}) \times \sigma(P) = 4 \times 703{,}428$	Number of filled elements: 430 883 950
	$\mathbb{A}_{L\rho}$	$\sum_{i=1}^{3} \sigma(\mathcal{E}_{l_i}) \times \sigma(\mathcal{E}_S) = 16 \times 1991$	Number of filled elements: 2263
System adjacency matrix Section 5.9 on page 152	$\mathbb{A}_{SC}$	$\left[\sum_{i=1}^{3} \sigma(\mathcal{E}_{l_i}) + \sigma(R_P)\sigma(P) + \sigma(R)\right] \times \left[\sum_{i=1}^{3} \sigma(\mathcal{E}_{l_i}) + \sigma(R_P)\sigma(P) + \sigma(R)\right]$	System Capabilities: 16,625
	$\widetilde{\mathbb{A}}_{SC}$	$\left[\sum_{i=1}^{3} \sigma(\mathcal{E}_{l_i}) + \sigma(\mathcal{E}_S) + \sigma(Q)\right] \times \left[\sum_{i=1}^{3} \sigma(\mathcal{E}_{l_i}) + \sigma(\mathcal{E}_S) + \sigma(Q)\right]$	System Capabilities: 16,625

model. The structure of an LFES is modeled using capabilities, as processes mapped onto resources, and the hetero-functional adjacency matrix, as a definition of system sequence. The sets of processes and resources are defined as mutually exclusive and collectively exhaustive, which ensures an isomorphic representation of the physical structure.

The controller model consists of the controller adjacency matrix, as a social network, and a controller agency matrix, which defines the cyber-physical interfaces with the physical structure. The set of cyber-resources is a subset of the system resources, and is therefore also mutually exclusive and collectively exhaustive within the system boundary. Additionally, the set of processes includes the decision algorithms P_Q. Consequently, both the cyber-resources and the cyber-processes are mutually exclusive, ensuring an isomorphic representation of the control structure.

Finally, the operand behavior describes the services of the system. In this chapter, both discrete and continuous states have been modeled with the service model. Consequently, the service model facilitates the representation of a wide variety of services, which has been demonstrated in the literature. Chapter 6 provides a more detailed discussion of the wide variety of application areas in which hetero-functional graph theory was applied.

Throughout Chapters 4 and 5, the work assesses the need for formal constructs to completely represent the conceptualization of the system. Additionally, it assesses if the mapping between the formal constructs and the system conceptualization is one-to-one. These assessments emphasize the ontological foundations of hetero-functional graph theory.

5.10.2 Comparison with Multi-layer Networks

Beyond the ontological strengths of hetero-functional graph theory, its application to interdependent smart city infrastructures fulfills a pressing theoretical need. The theoretical versatility required to model smart city infrastructure systems quantitatively has not been realized previously. The multi-layer networks literature imposes constraints on the systems they model, as discussed by the extensive review of Kivela et al. [14]. Naturally, such constraints, as discussed in Chapter 2 on Page 13, limit the ability of multi-layer networks to model an arbitrary number of arbitrarily coupled smart city infrastructures. Hetero-functional graph theory does not impose these constraints. For each of the eight constraints identified by Kivela et al., an example in the Trimetrica test case is provided to show that hetero-functional graph theory is not similarly limited. To facilitate the discussion, the three infrastructure systems in Trimetrica are considered to be the three layers of the multi-layer network.

In *Constraint 1*, some "multi-layer networks" require all layers to have vertically aligned nodes. Hetero-functional graph theory, however, does not impose such a requirement. For example, Trimetrica contains substations that only appear in the electric power system layer. Even if resources are geographically in the same location, they are not required to be connected. For example, Resource 190 and Resource 269 represent a substation and an intersection, respectively. These have the same GPS location, but are not connected.

In *Constraint 2*, some "multi-layer networks" require disjoint layers. Hetero-functional graph theory, however, does not impose such a requirement. Trimetrica has resources that are part of more than a single infrastructure system. For example, houses with EV chargers appear in the water distribution system, electric power system, and the electrified transportation system.

In *Constraint 3*, some "multi-layer networks" require the same number of nodes in each layer. Hetero-functional graph theory, however, allows for an arbitrary number of nodes in each layer. For example, in the Trimetrica test case, the electric power system has 201 nodes, whereas the electrified transportation system has 169 nodes.

In *Constraint 4*, some "multi-layer networks" require exclusively "vertical"[10] inter-layer couplings. Hetero-functional graph theory, however, allows for arbitrary couplings between layers. For example, resource 6 is a house with an EV charger. This house consumes hot water which draws power that arrives by power line. Consequently, there is a coupling between the power line's transport power capability and the house's hot water consumption capability, which is not a vertical coupling.

In *Constraint 5*, some "multi-layer networks" require that all nodes in a given layer have identical couplings to nodes in another layer. Hetero-functional graph theory, however, is able to represent multiple distinct couplings to nodes in another layer. For example, in Trimetrica, Resource 1 couples the electric power system and the water distribution system as it is a water treatment facility with an EV charger that consumes electric power by virtue of its "treat water" process. However, Resource 6 also couples the electric power system and the water distribution system as it is a house with an EV charger that consumes electric power by virtue of its "consume hot water" process. The nature of these couplings is fundamentally different, as the processes are distinct.

In *Constraint 6*, some "multi-layer networks" require that each node is connected to all of its counterparts in other layers. Hetero-functional graph theory, however, does not require that each node is connected to all of its counterparts. For example, in Trimetrica, Resource 6 is defined as a House with an EV charger. The house has multiple, unrelated capabilities, such as "park EV at House 6" and "consume cold water at House 6." These capabilities are not sequentially coupled, as no sequence has been defined that couples the two capabilities (as follows from Figure 5.11 on Page 112). Consequently, even if the resource is part of multiple layers, the components of that resource are not necessarily coupled.

[10]Kivela et al. [14] refer to this constraint as "diagonal" couplings. This work adopts the term vertical to more closely reflect the depiction in Figure 2.1.

In *Constraint 7*, some "multi-layer networks" limit the number of layers to two. Hetero-functional graph theory, however, facilitates the modeling of an arbitrary number of layers. For example, the Trimetrica test case consists of three infrastructure systems.

In *Constraint 8*, some "multi-layer networks" require that each layer have no more than one aspect. Hetero-functional graph theory, however, allows for an arbitrary number of aspects in a layer. For example, the electrified transportation system has both conventional roads and electrified roads. The electrified roads have fundamentally different characteristics than the conventional road, but they are both represented within the transportation system.

References

1. Allan, R., Billinton, R., Abdel-Gawad, N. (1986). The IEEE reliability test system - extensions to and evaluation of the generating system. *Power Engineering Review, IEEE, PER-6*(11), 24–24.
2. Allan, D. F., Mezher, T., & Farid, A. M. (2016). Enhanced electric vehicle adoption scenarios for Abu Dhabi Road transportation. In *UAE Graduate Students Research Conference*, Al Ain, pp. 1–2. [Online]. Available: http://engineering.dartmouth.edu/liines/resources/Conferences/TES-C57.pdf
3. Center for Water Systems. (2006). Anytown water distribution network. [Online]. Available: http://emps.exeter.ac.uk/engineering/research/cws/resources/benchmarks/expansion/anytown.php
4. Committee on Increasing National Resilience to Hazards and Disasters and Committee on Science Engineering and Public Policy. (2012). *Disaster resilience: A national imperative*. Washington, DC: The National Academies Press.
5. David, R., & Alla, H. (2010). *Discrete, continuous, and hybrid Petri nets*. Berlin, Heidelberg: Springer.
6. D'Agostino, G., & Scala, A. (2014). *Networks of networks: The last frontier of complexity* (Vol. 340). Berlin, Heidelberg: Springer.
7. Department of Homeland Security. (2013). National infrastructure protection plan: Partnering for critical infrastructure security and resilience. Department of Homeland Security, Washington, D.C. United states, Tech. Rep.
8. de Oliveira de Jesus, P. M. (2007). Remuneration of distributed generation: A holistic approach. Ph.D. dissertation, Faculdade de Engharia Universidade de Porto.
9. Farid, A. M. (2015). Symmetrica: Test case for transportation electrification research. *Infrastructure Complexity*, *2*(9), 1–10. [Online]. Available: http://dx.doi.org/10.1186/s40551-015-0012-9
10. Farid, A. M. (2015). LIINES datasets repository. In *Laboratory for Intelligent Integrated Networks of Engineering Systems*, Hanover, NH, p. 1. [Online]. Available: https://engineering.dartmouth.edu/liines/Datasets/index.html
11. Farid, A. M. (2016). A hybrid dynamic system model for multi-modal transportation electrification. *IEEE Transactions on Control System Technology*, *PP*(99), 1–12. [Online]. Available: http://dx.doi.org/10.1109/TCST.2016.2579602
12. Grigg, C., Wong, P., Albrecht, P., Allan, R., Bhavaraju, M., Billinton, R., et al. (1999). The IEEE reliability test system-1996. A report prepared by the reliability test system task force of the application of probability methods subcommittee. *IEEE Transactions on Power Systems*, *14*(3), 1010–1020.

13. Haimes, Y. Y., Crowther, K., & Horowitz, B. M. (2008). Homeland security preparedness: Balancing protection with resilience in emergent systems. *Systems Engineering, 11*(4), 287–308.
14. Kivelä, M., Arenas, A., Barthelemy, M., Gleeson, J. P., Moreno, Y., Porter, M. A. (2014). Multilayer networks. *Journal of Complex Networks, 2*(3), 203–271.
15. Ramirez-Rosado, I. J., & Bernal-Agustin, J. L. (1998). Genetic algorithms applied to the design of large power distribution systems. *IEEE Transactions on Power Systems, 13*(2), 696–703.
16. Schoonenberg, W. C., & Farid, A. M. (2017). Modeling smart cities with hetero-functional graph theory. In *2017 IEEE International Conference on Systems, Man, and Cybernetics (SMC2017), Intelligent Industrial System Special Session* (Vol. 1, no. 1, pp. 1–10).
17. Subcommittee, P. (1979). IEEE reliability test system. *IEEE Transactions on Power Apparatus and Systems, PAS-98*(6), 2047–2054.
18. The White House Office of the Press Secretary (2013). Presidential policy directive: Critical infrastructure security and resilience (PPD-21). The White House, Washington, D.C. United states, Tech. Rep. [Online]. Available: http://www.whitehouse.gov/the-press-office/2013/02/12/presidential-policy-directive-critical-infrastructure-security-and-resil
19. Treiber, M., & Kesting, A. (2010). An open-source microscopic traffic simulator. *Intelligent Transportation Systems Magazine, IEEE, 2*(3), 6–13.
20. van der Wardt, T. J., & Farid, A. M. (2017). A hybrid dynamic system assessment methodology for multi-modal transportation-electrification. *Energies, 10*(5), 653. [Online]. Available: http://dx.doi.org/10.3390/en10050653

Chapter 6
Applications of Hetero-functional Graph Theory

As mentioned in the book's preface and later in Chapter 4, hetero-functional graph theory can be viewed as an intellectual fusion of model-based systems engineering and network science. Much like these two fields, it has also been applied to numerous application domains. This chapter serves to highlight this breadth of applications. In all, six application domains are addressed:

1. Mass-Customized Production Systems
2. Multi-Modal Transportation Systems
3. Electric Power Systems
4. Multi-Modal Electrified Transportation Systems
5. Microgrid-Enabled Production Systems
6. Personalized Healthcare Delivery Systems

Throughout these works, there have been three consistent themes of methodological contributions. First, in all cases, the complete LFES meta-architecture shown in Figure 6.1 was instantiated to create an ***application domain-specific reference architecture*** following the logic presented in Figure 3.4. Second, such a reference architecture serves the derivation of ***measures of life-cycle properties***. Finally, equations of motion are imposed on the domain-specific reference architecture to develop ***dynamic simulation models*** for in-depth ***systems analysis***. These three themes of methodological contributions are also highlighted as each application domain is discussed.

6.1 Mass-Customized Production Systems

Hetero-functional graph theory originated from the mass customized production system literature. There, the need to compete in dynamic marketplaces with product variants of increasingly short product life-cycle drove mass-customized production

W. C. H. Schoonenberg et al., *A Hetero-functional Graph Theory for Modeling Interdependent Smart City Infrastructure*,
https://doi.org/10.1007/978-3-319-99301-0_6

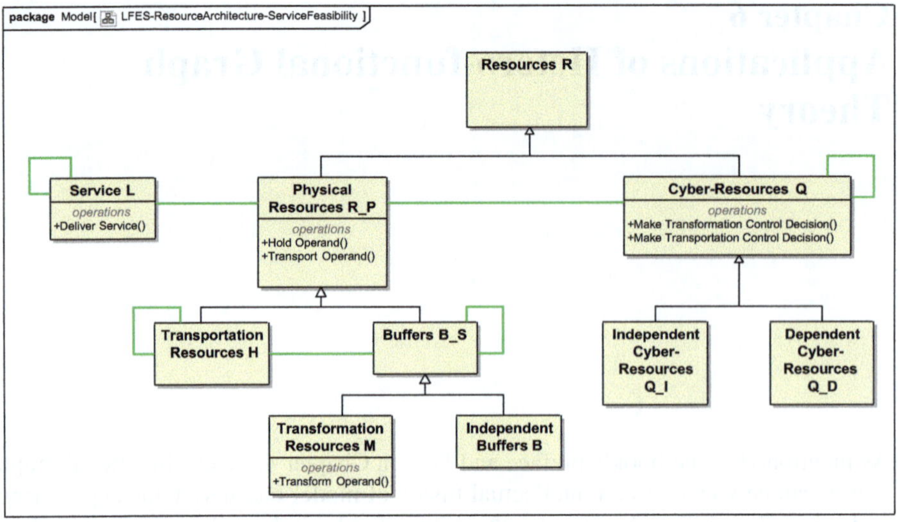

Fig. 6.1 A SysML Block Diagram: The meta-architecture of the system form of an LFES with cyber-resources and the service model

systems to explicitly foster ***reconfigurability*** as a life-cycle property of their integrated automation solutions. To that end, Prof. Farid's doctoral dissertation *reconfigurability measurement in automated manufacturing systems* (2007) specifically developed quantitative measures of reconfigurability [4, 8, 17]. There it was found that reconfigurability depended on two concepts: ***reconfiguration ease*** and ***reconfiguration potential***. The former used a production ***design structure matrix*** as a type of graph [21–23] that captured the complexity of interfaces between production system resources. However, it was soon recognized that such methods were insufficient to address why distributed manufacturing control systems were so much more easy to reconfigure than central manufacturing control systems. In that regard, it became necessary to distinguish between a production design structure matrix defined along resource boundaries and one defined along production capabilities (i.e., degrees of freedom) [4, 5]. Meanwhile, the relatively new concept of reconfiguration potential was introduced to differentiate between production resources that had different types or numbers of capabilities [4, 6, 20, 24]. The two concepts of reconfiguration potential and ease were then used to develop measures for the key (life-cycle) characteristics of reconfigurability: modularity, customization, integrability, and convertibility [4, 8, 17].

This initial application of hetero-functional graph theory to the reconfigurability of mass-customized production systems brought about further developments in system life-cycle properties that are of direct relevance to the network science literature. In their review on multi-layer networks, Kivela et al. [38] write: *"Research on generalizing monoplex-network concepts such as degree, transitivity, centrality and diffusion is only in its infancy. We also expect that it will be necessary to define many concepts that are intrinsic to multi-layer networks. … Walks and paths – and their lengths – are important concepts in both graph theory and network science. The ability to define generalizations of such concepts for multi-layer networks yields natural extensions of many other measures for monoplex networks, including graph distance, connected components, betweenness centralities, random walks, communicability and clustering coefficients."* To that effect, hetero-functional graph theory was used to develop the concept of ***production paths*** [7]. Unlike paths in traditional graphs, production paths explicitly consider the feasibility of a product as it proceeds through its various stages of a mass-customized production system. This work led to the development of new hetero-functional graph ***resilience measures*** [10, 11, 14] shortly thereafter. Furthermore, the concept of measuring the effort of a reconfiguration process [18] was introduced analogously to the concepts of graph percolation [3, 39] and temporal networks [28]. This early work addressed ***dynamic structure*** of large flexible engineering systems [18]. Later on, hetero-functional graph theory was extended with Petri nets to address the ***dynamic behavior*** of these mass-customized production systems. Finally, these works directly informed the quantitative development of ***distributed multi-agent (control) system architectures*** that intrinsically exhibit reconfigurability and resilience [12, 25, 26]. In that regard, hetero-functional graph theory explicitly demonstrated its ability to address ***cyber-physical*** systems.

6.2 Multi-Modal Transportation Systems

The work on mass-customized production systems spawned research on multi-modal transportation systems. In many ways, transportation systems can be viewed as a *degenerate* production system without any transformation processes. Consequently, many of the developments described in the previous subsection were customized for multi-modal transportation systems as well. From an architectural perspective, hetero-functional graph theory treats multi-modal transportation system with ease. Unlike other works which treat multiple modes of transportation as different *aspects* on a single graph structure, hetero-functional graphs distinguish between buses and trains using different types of system resources [1, 2, 43]. Furthermore, holding processes can be introduced so as to distinguish transportation processes that have different operands or hold those operands in a different way. In these works, ***passenger itineraries*** were enumerated as an analogous concept to production paths.

6.3 Electric Power Systems

Shortly thereafter, it became clear that hetero-functional graph theory would have cross-domain relevance. One pending question was its applicability to large flexible engineering systems with continuous-time rather than discrete-event equations of motion. In that regard, and from an architectural perspective, the theory was applied to calculate the system concept and hetero-functional adjacency matrix of an electric power system [9, 13]. The latter of these explicitly showed how hetero-functional graph theory models can be used to derive two well-established dynamic models of electric power systems: ***the power flow analysis model and the transient stability model*** [13]. Interestingly, the hetero-functional graph derivation explicitly accommodates power lines of unequal impedance in opposite directions whereas the traditional graph theoretic method does not. Consequently, the hetero-functional graph method is better suited to future electric power systems that have made extensive use of FACTS devices [27]. Finally, these works, much like their analogies in mass-customized production systems, informed the quantitative development of ***distributed multi-agent control system architectures*** that intrinsically enable ***resilient electric power systems***.

6.4 Multi-Modal Electrified Transportation Systems

With the application of hetero-functional graph theory to both multi-modal transportation systems and electric power systems firmly established, it is only natural to integrate these works to address multi-modal electrified transportation systems. Again, the LFES meta-architecture was instantiated to create the associated domain-specific reference architecture [15]. Once in place, dynamic equations of motion called ***device models*** were added to each system capability (i.e., degree of freedom) in the architecture to produce a dynamic simulation model [15, 16, 44]. Unlike previous works that had used hetero-functional graph theory to re-derive well-known equations of motion for single infrastructures, these works were now showing that hetero-functional graph theory could be used to develop dynamic models of arbitrarily coupled interdependent infrastructure systems. Multi-modal electrified transportation systems are particularly interesting because they are described by a ***hybrid dynamic model*** in which the transportation system is described a ***discrete-event Petri net*** and the electric power system is described by continuous-time power flow equations. Furthermore, such a model showed that it can be used to distinguish between fundamentally different ways of coupling multiple infrastructure. For example, one work [16] compared the performance of ***online electric vehicles*** that charge wirelessly *while* driving and conventional electric vehicles that charge by wire while parking. From a traditional graph theory perspective, the latter represents a node-to-node coupling while the former is a node-to-edge coupling that has yet to appear in the multi-layer network literature. Finally, the development of such models enable the development of robust assessment methodologies that rigorously describe data requirements, analysis steps, and performance measures [42].

6.5 Microgrid-Enabled Production Systems

Hetero-functional graph theory was subsequently extended to microgrid-enabled production systems [40, 41]. In this case, the ***discrete-event Petri net*** dynamics of a mass-customized production system were coupled to ***continuous-time power flow dynamics*** of a microgrid in a ***hybrid dynamic model***. The intention behind this work was to demonstrate how hetero-functional graph theory can be used to address the pressing energy management challenges that many industrial facilities face as they make efforts to decarbonize. Again, hetero-functional graph theory was integral to the development of the associated domain-specific reference architecture and the subsequent addition of dynamic equations of motion.

6.6 Personalized Healthcare Delivery Systems

Perhaps the greatest demonstration of the versatility of hetero-functional graph theory was in the domain of personalized healthcare delivery systems. The first such works addressed the healthcare delivery system at the ***planning*** time scale where the architecture of the healthcare delivery system can change from year to year. In the ***first longitudinal study of its kind***, hetero-functional graph theory was used to track the ***evolution of the Abu Dhabi healthcare labor pool from 1967–2012*** [19, 29, 30]. The study viewed the evolution of the Abu Dhabi Healthcare system concept as a temporal bipartite network that experiences targeted human resource management "reconfiguration processes." It showed that while the Emirate had done much to successfully broaden and deepen its healthcare labor pool with respect to healthcare specialization and geographic representation, it also had much left to do to address per capita service availability and human resource capital attrition.

This application of hetero-functional graph theory to personalized healthcare delivery systems at the planning time scale inspired its application at the time scale of ***clinical operations***. Here, it was recognized that the unprecedented chronic disease burden now facing healthcare delivery systems around the world motivates a shift ***from reductionist to system thinking***. Consequently, new systems tools are needed in the practice of clinical medicine [31, 36]. At this point, hetero-functional graph theory was applied to develop a domain-specific ***reference architecture for personalized healthcare delivery systems*** [32, 34, 37]. Unlike industrial engineering and operations research works that only consider the evolution of the healthcare delivery system [31, 36], this reference architecture explicitly models the ***dynamic and stochastic evolution of an individual's health state*** as a ***fuzzy Petri net***. Such a ***health net*** model is of direct relevance to understanding the evolution of chronic disease in smart cities [33]. Finally, this personalized healthcare delivery system reference architecture was developed into a full discrete-event dynamic model in which the evolution of the healthcare delivery system could be studied simultaneously with the stochastic evolution of an individual's health state [35]. Consequently, system-wide healthcare costs could, for the first time, be directly tied to healthcare outcomes.

References

1. Baca, E. E. S., & Farid, A.M. (2013) An axiomatic design approach to reconfigurable transportation systems planning and operations (invited paper). In *DCEE 2013: 2nd International Workshop on Design in Civil & Environmental Engineering* (pp. 22–29). Worcester. http://engineering.dartmouth.edu/liines/resources/Conferences/TES-C11.pdf
2. Baca, E. E. S., Farid, A. M., Tsai, I.-T., & Viswanath, A. (2013) An axiomatic design approach to passenger itinerary enumeration in reconfigurable transportation systems. In *Proceedings of ICAD2013 The Seventh International Conference on Axiomatic Design* (vol. PP, no. 99, pp. 1–10). Worcester. http://engineering.dartmouth.edu/liines/resources/Conferences/TES-C23.pdf
3. Callaway, D. S., Newman, M. E. J., Strogatz, S. H., & D. J. Watts (2000) Network robustness and fragility: Percolation on random graphs. *Physical Review Letters, 85*, 5468–5471. http://link.aps.org/doi/10.1103/PhysRevLett.85.5468
4. Farid, A. M. (2007) Reconfigurability measurement in automated manufacturing systems. Ph.D. Dissertation, University of Cambridge Engineering Department Institute for Manufacturing. http://engineering.dartmouth.edu/liines/resources/Theses/IEM-TP00.pdf
5. Farid, A. M. (2008) Facilitating ease of system reconfiguration through measures of manufacturing modularity. *Proceedings of the Institution of Mechanical Engineers, Part B (Journal of Engineering Manufacture) – invited paper, 222*(B10), 1275–1288. http://dx.doi.org/10.1243/09544054JEM1055
6. Farid, A. M. (2008) Product degrees of freedom as manufacturing system reconfiguration potential measures. *International Transactions on Systems Science and Applications – invited paper, 4*(3), 227–242. http://engineering.dartmouth.edu/liines/resources/Journals/IEM-J04.pdf
7. Farid, A. M. (2013) An axiomatic design approach to non-assembled production path enumeration in reconfigurable manufacturing systems. In *2013 IEEE International Conference on Systems Man and Cybernetics* (pp. 1–8). Manchester. http://dx.doi.org/10.1109/SMC.2013.659
8. Farid, A. M. (2014) Axiomatic design and design structure matrix measures for reconfigurability and its key characteristics in automated manufacturing systems. In *International Conference on Axiomatic Design* (pp. 1–8). Caparica. http://engineering.dartmouth.edu/liines/resources/Conferences/IES-C39.pdf
9. Farid, A. M. (2014) Multi-agent system design principles for resilient operation of future power systems. In *IEEE International Workshop on Intelligent Energy Systems* (pp. 1–7). San Diego. http://engineering.dartmouth.edu/liines/resources/Conferences/SPG-C42.pdf
10. Farid, A. M. (2014) Static resilience of large flexible engineering systems: Part I – axiomatic design model. In *4th International Engineering Systems Symposium* (pp. 1–8). Hoboken, NJ: Stevens Institute of Technology. http://engineering.dartmouth.edu/liines/resources/Conferences/IES-C37.pdf
11. Farid, A. M. (2014) Static resilience of large flexible engineering systems: Part II – axiomatic design measures. In *4th International Engineering Systems Symposium* (pp. 1–8). Hoboken, N.J.: Stevens Institute of Technology. http://engineering.dartmouth.edu/liines/resources/Conferences/IES-C38.pdf

12. Farid, A. M. (2015) Designing multi-agent systems for resilient engineering systems (invited paper). In *Holomas 2015 7th International Conference on Industrial Applications of Holonic and Multi-Agent Systems* (pp. 1–6). Valencia. http://dx.doi.org/10.1007/978-3-319-22867-9_1
13. Farid, A. M. (2015) Multi-agent system design principles for resilient coordination and control of future power systems. *Intelligent Industrial Systems, 1*(3), 255–269. http://dx.doi.org/10.1007/s40903-015-0013-x
14. Farid, A. M. (2015) Static resilience of large flexible engineering systems: Axiomatic design model and measures. *IEEE Systems Journal, PP*(99), 1–12. http://dx.doi.org/10.1109/JSYST.2015.2428284
15. Farid, A. M. (2016) A hybrid dynamic system model for multi-modal transportation electrification. *IEEE Transactions on Control System Technology, PP*(99), 1–12. http://dx.doi.org/10.1109/TCST.2016.2579602
16. Farid, A. M. (2017) Electrified transportation system performance: Conventional vs. online electric vehicles. In N. P. Suh & D. H. Cho (Eds.), *The on-line electric vehicle: Wireless electric ground transportation systems* (Chap. 20, pp. 279–313). Berlin: Springer. http://engineering.dartmouth.edu/liines/resources/Books/TES-BC05.pdf
17. Farid, A. M. (2017) Measures of reconfigurability and its key characteristics in intelligent manufacturing systems. *Journal of Intelligent Manufacturing, 28*(2), 353–369. http://dx.doi.org/10.1007/s10845-014-0983-7
18. Farid, A. M., & Covanich, W. (2008) Measuring the effort of a reconfiguration process. In *IEEE International Conference on Emerging Technologies and Factory Automation, 2008. ETFA 2008* (pp. 1137–1144). Hamburg. http://dx.doi.org/10.1109/ETFA.2008.4638540
19. Farid, A. M., & Khayal, I. S. (2013) Axiomatic design based volatility assessment of the abu dhabi healthcare labor market: Part I-theory. In *Proceedings of the ICAD 2013: The Seventh International Conference on Axiomatic Design* (pp. 1–8). Worcester, MA. http://engineering.dartmouth.edu/liines/resources/Conferences/IES-C21.pdf
20. Farid, A. M., & McFarlane, D. C. (2006) A development of degrees of freedom for manufacturing systems. In *IMS'2006: 5th International Symposium on Intelligent Manufacturing Systems: Agents and Virtual Worlds* (pp. 1–6). Sakarya. http://engineering.dartmouth.edu/liines/resources/Conferences/IEM-C02.pdf
21. Farid, A. M., & McFarlane, D. C. (2006) An approach to the application of the design structure matrix for assessing reconfigurability of distributed manufacturing systems. In *Proceedings of the IEEE Workshop on Distributed Intelligent Systems: Collective Intelligence and Its Applications (DIS'06)* (pp. 1–6). Prague. http://dx.doi.org/10.1109/DIS.2006.10
22. Farid, A. M., & McFarlane, D. C. (2006) A tool for assessing reconfigurability of distributed manufacturing systems. In *12th IFAC Symposium on Information Control Problems in Manufacturing* (Vol. 12, no. PART 1, pp. 1–6). Saint Etienne. http://engineering.dartmouth.edu/liines/resources/Conferences/IEM-C03.pdf
23. Farid, A. M., & McFarlane, D. C. (2007) A design structure matrix based method for reconfigurability measurement of distributed manufacturing systems. *International Journal of Intelligent Control and Systems Special Issue – Invited Paper, 12*(2), 118–129. http://engineering.dartmouth.edu/liines/resources/Journals/IEM-J02.pdf
24. Farid, A. M., & McFarlane, D. C. (2008) Production degrees of freedom as manufacturing system reconfiguration potential measures. *Proceedings of the Institution of Mechanical Engineers, Part B (Journal of Engineering Manufacture) – invited paper, 222*(B10), 1301–1314. http://dx.doi.org/10.1243/09544054JEM1056
25. Farid, A. M., & Ribeiro, L. (2014) An axiomatic design of a multi-agent reconfigurable manufacturing system architecture. In *International Conference on Axiomatic Design* (pp. 1–8). Lisbon. http://engineering.dartmouth.edu/liines/resources/Conferences/IEM-C41.pdf
26. Farid, A. M., & Ribeiro, L. (2015) An axiomatic design of a multi-agent reconfigurable mechatronic system architecture. *IEEE Transactions on Industrial Informatics, 11*(5), 1142–1155. http://dx.doi.org/10.1109/TII.2015.2470528

27. Hingorani, N.G., & Gyugyi, L. (2000) *Understanding FACTS : Concepts and technology of flexible AC transmission systems*. New York: IEEE Press. http://ieeexplore.ieee.org/xpl/bkabstractplus.jsp?bkn=5264253
28. Holme, P., & Saramäki, J. (2012) Temporal networks. *Physics Reports: Review Section of Physics Letters, 519*(3), 97–125
29. Khayal, I. S., & Farid, A. M. (2013) axiomatic design based volatility assessment of the abu dhabi healthcare labor market: Part II – case study. In *Proceedings of the ICAD 2013: The Seventh International Conference on Axiomatic Design* (pp. 1–8). Worcester, MA. http://engineering.dartmouth.edu/liines/resources/Conferences/IES-C20.pdf
30. Khayal, I. S., & Farid, A. M. (2015) Axiomatic design based volatility assessment of the Abu Dhabi Healthcare labor market. *Journal of Enterprise Transformation, 5*(3), 162–191. http://dx.doi.org/10.1080/19488289.2015.1056449
31. Khayal, I. S., & Farid, A. M. (2016) The need for systems tools in the practice of clinical medicine. *Systems Engineering, 20*(1), 3–20. http://dx.doi.org/10.1002/sys.21374
32. Khayal, I., & Farid, A. (2017) An architecture for a cyber-physical healthcare delivery system with human agents. In *Proceedings of the 2017 IEEE International Summer School on Smart Cities (IEEE S3C)*
33. Khayal, I., & Farid, A. (2017) Designing smart cities for citizen health & well-being. In *Proceedings of the 2017 IEEE International Summer School on Smart Cities (IEEE S3C)*
34. Khayal, I. S. & Farid, A. M. (2017) Architecting a system model for personalized healthcare delivery and managed individual health outcomes. *Complexity, 1*(1), 1–25. https://doi.org/10.1155/2018/8457231
35. Khayal, I., & Farid, A. M. (2017) A dynamic model for a cyber-physcal healthcare delivery system with human agents. In *2017 IEEE International Conference on Systems, Man, and Cybernetics (SMC2017), Intelligent Industrial System Special Session*. Banff
36. Khayal, I., & Farid, A. M. (2017) The application of model-based systems engineering to the practice of clinical medicine. In *Proceedings of the 11th Annual IEEE International Systems Conference*. Montreal, QC:
37. Khayal, I., & Farid, A. (2018) Designing patient-oriented healthcare services as systems of systems. In *2018 IEEE Systems of Systems Conference*. Paris
38. Kivelä, M., Arenas, A., Barthelemy, M., Gleeson, J. P., Moreno, Y., & Porter, M. A. (2014) Multilayer networks. *Journal of complex networks, 2*(3), 203–271
39. Newman, M. E. (2003) The structure and function of complex networks. *SIAM Review, 45*(2), 167–256
40. Schoonenberg, W. C., & Farid, A. M. (2015) A dynamic production model for industrial systems energy management. In *2015 IEEE International Conference on Systems Man and Cybernetics* (pp. 1–7). Hong Kong. http://dx.doi.org/10.1109/SMC.2015.14
41. Schoonenberg, W. C., & Farid, A. M. (2017) A dynamic model for the energy management of microgrid-enabled production systems. *Journal of Cleaner Producation, 1*(1), 1–10
42. van der Wardt, T. J., & Farid, A. M. (2017) A hybrid dynamic system assessment methodology for multi-modal transportation-electrification. *Energies, 10*(5), 653. http://dx.doi.org/10.3390/en10050653
43. Viswanath, A., Baca, E. E. S., & Farid, A. M. (2014) An axiomatic design approach to passenger itinerary enumeration in reconfigurable transportation systems. *IEEE Transactions on Intelligent Transportation Systems, 15*(3), 915–924. http://dx.doi.org/10.1109/TITS.2013.2293340
44. Viswanath, A., Farid, A. M. (2013) A hybrid dynamic system model for the assessment of transportation electrification. In *5th MIT-MI Joint Workshop on the Reliability of Power System Operation & Control in the Presence of Increasing Penetration of Variable Energy Sources* (pp. 1–28). Abu Dhabi

Chapter 7
Conclusion and Future Work

7.1 Conclusion

This book has advanced a hetero-functional graph theory to analyze the structure of interdependent smart city infrastructures. In that regard, it simultaneously fulfills two pressing needs: one practical and one theoretical. Twenty first century smart cities represent a tremendous practical challenge. They are now ever-more connected with other smart cities around the globe, they are growing to become mega-cities with 10 million inhabitants each, and they are infused with ubiquitous information technology. The looming challenge of global climate change further raises questions of these cities' sustainability and resilience, in the present, and in years to come. Consequently, the unprecedented need for big data analytics to address these challenges is met, quite fortunately, with an equally unprecedented availability of ubiquitous data. And yet, as smart cities become more crowded, systems theory recognizes that its various infrastructure systems will become increasingly interdependent. Consequently, the analytical techniques, developed primarily for single infrastructure systems, will have to develop with an appropriate theory for interdependent infrastructure systems.

Chapter 2 deeply investigated the need for hetero-functional graph theory in the context of the existing literature on multi-layer network theory. It showed by virtue of a simple example that the eight modeling constraints in the multi-layer network literature present a practical limitation in modeling multiple arbitrarily connected infrastructure systems.

In the context of this literature gap, Chapter 3 introduced ontological science, model-based systems engineering and graph theory as a theoretical foundation for hetero-functional graph theory. The ontological analysis in Chapter 3 demonstrated that, with respect to a domain conceptualization of interdependent infrastructure systems, multi-layer networks maintain soundness and laconicity. However, they lack completeness because they do not address system function and they lack

W. C. H. Schoonenberg et al., *A Hetero-functional Graph Theory for Modeling Interdependent Smart City Infrastructure*,
https://doi.org/10.1007/978-3-319-99301-0_7

lucidity because the representations of nodes and edges are overloaded. Therefore, model-based systems engineering was introduced as a more ontologically appropriate domain conceptualization for interdependent infrastructure systems.

Chapter 4 provided an exposition of hetero-functional graph theory as an intellectual fusion of model-based systems engineering and network science. It is presented in terms of its constituent mathematical models and how they relate to their counterparts in SysML. These models are: (1) the System Concept, (2) the Hetero-functional Adjacency Matrix, (3) the Controller Agency Matrix, (4) the Controller Adjacency Matrix, (5) the Service as Operand Behavior, (6) the Service Feasibility Matrix, and (7) the System Adjacency matrix. The first two models are assumed to be universal and apply to all types of engineering systems. They form the structural model. Because there is an explicit differentiation of system processes in the system concept, there is a basis upon which networks with unlike function can be combined into a single mathematical model. Furthermore, and unlike traditional applications of graph theory, where nodes and edges are both elements of system form, the nodes of the hetero-functional adjacency matrix represent system capabilities and the edges represent the logical sequences between them. Moving on, Models 3 and 4 constitute the system control model. Together, they differentiate systems based upon the structure of their control and decision-making. Models 5 and 6 constitute the service model. They differentiate systems based upon the behavior of their operands. These six models are then ultimately coupled together for holistic analysis of cyber-physical engineering systems. The structural model and the control model have cyber-physical resource interfaces. Similarly, the structural model and the service model have a structure-service coupling. When integrated together, these models constitute the final product of hetero-functional graph theory: the System Adjacency Matrix. Throughout Chapter 4, multiple *ontological remarks* were added to relate the mathematical models back to the ontological concepts presented in Chapters 2 and 3. To ground the discussion, Chapter 4 demonstrates the entirety of the theory on a simple example. To the systems engineering community, hetero-functional graph theory is a quantification of many of the structural concepts found in model-based systems engineering languages like SysML. To the network science community, it presents a new view as how to construct graphs with fundamentally different meaning and insight.

With the theory in place, Chapter 5 introduced the "Trimetrica" test case as an interdependent smart city infrastructure system and then proceeded to use hetero-functional graph theory to describe the test case as a single mathematical model. An incremental approach was taken to show how each of the first six models described above can be used to develop a full system adjacency matrix piece-by-piece. Along the way, the graph theoretic concepts were extensively visualized for greater intuition and clarity. The heterogeneity found in the "Trimetrica" test case demonstrated the versatility of hetero-functional graph theory in modeling an arbitrary number of arbitrarily coupled interdependent infrastructures. This versatility is notable because the interactions between multiple interdependent infrastructures has permutation complexity. Chapter 5 showed that whereas two infrastructure systems can interact in $P(2,2) = 2!/0! = 2$ ways, three infrastructure systems can interact in $P(3,2) = 3!/1! = 6$ ways. To further demonstrate the versatility of the hetero-functional graph theory, Chapter 5 showed how the hetero-functional graph model overcomes all of the modeling and ontological limitations found in the multi-layer network literature.

Finally, Chapter 6 highlighted the many already-existing applications of hetero-functional graph theory. These include:

1. mass-customized production systems,
2. multi-modal transportation systems,
3. electric power systems,
4. multi-modal electrified transportation systems,
5. microgrid-enabled production systems, and
6. personalized healthcare delivery systems.

Throughout these works, there have been three consistent themes of methodological contributions. First, in all cases, the complete LFES meta-architecture was instantiated to create an application domain-specific reference architecture. Second, such a reference architecture serves the derivation of measures of life-cycle properties. Finally, equations of motion are imposed on the domain-specific reference architecture to develop dynamic simulation models for in-depth systems analysis.

7.2 Future Work

From the above section, it is clear that hetero-functional graph theory has already been used for a wide variety of practical and theoretical applications. However, in both practical and theoretical dimensions, the integration of networks of systems is only in its infancy. Straightforward practical future directions include its application to other infrastructure systems of different type, size, and number. More theoretically, hetero-functional graph theory lends itself to provide new perspective measures of life-cycle properties and network characteristics with special emphasis on the coupling of multiple disciplines. Finally, and as related to previous work on dynamic systems, the dynamics of smart cities is a clear research direction for future exploration. In this regard, the integration of smart city infrastructure systems has the potential to facilitate holistic operation and realize synergistic sustainability and resilience benefits.

Appendix A
Representing a Four-Layer Network in Hetero-functional Graph Theory

Smart cities create the need for greater integration between multiple infrastructures. Here, it is important to note that the coupling between these multiple infrastructures may take on a wide variety of potential topologies. In order to facilitate the diversity of interdependent smart city infrastructure, a modeling language with great mathematical versatility is needed. To that end, this book develops hetero-functional graph theory. Chapter 2 introduces a four-layer network that violates the limitations imposed by the existing multi-layer network literature. This appendix discusses the four-layer network in more detail and presents the derivation of its hetero-functional graph representation. The goal of this appendix is to demonstrate that hetero-functional graph theory is able to represent a system that cannot be modeled with other multi-layer network methods. Furthermore, the model is ontologically sound, complete, lucid, and laconic. Such a model is achieved by following the theory of Chapter 4 closely. However, for a more detailed demonstration of hetero-functional graph theory, the reader is referred to Chapter 5 in the book in which the hetero-functional graph theory model of the Trimetrica test case is presented. This appendix proceeds as follows: first, it introduces the four-layer network. Section A.1 derives the system concept. Section A.2 calculates the hetero-functional adjacency matrix. Section A.3 calculates the controller agency matrix. Section A.4 derives the controller adjacency matrix. Section A.5 derives the service model for the four-layer network. Section A.6 calculates the service feasibility matrices. Finally, Section A.7 integrates the 6 hetero-functional graph theory models to construct the system adjacency matrix.

Figure A.1 presents the four-layer example network as used in Chapter 2. The *water distribution network* consists of three nodes and two edges. The water is supplied by the water treatment facility in Node v_9. This water is consumed as cold potable water in the house (Node v_8), and as hot water for the hot tub (Node v_{10}). The hot tub and the water treatment facility are both connected to the electric power system, because their processes require a supply of electric power.

W. C. H. Schoonenberg et al., *A Hetero-functional Graph Theory for Modeling Interdependent Smart City Infrastructure*,
https://doi.org/10.1007/978-3-319-99301-0

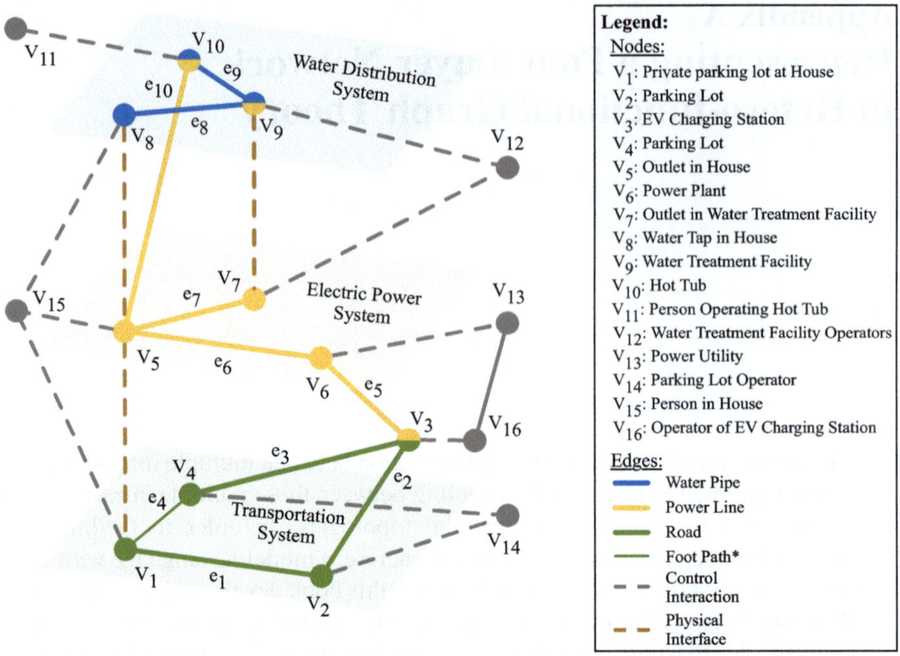

Fig. A.1 A Hypothetical Four-Layer Network: It represents transportation, electric power, and water distribution infrastructure with a superimposed cyber-control layer. *: The foot path is part of the Transportation System, but differs in modality from the other edges in the system and is represented with a thinner edge. Its two-dimensional representation is presented in Figure A.2

The *electric power system* consists of five nodes and four edges. The power plant in node v_6 supplies electric power for the electric power grid. The hot tub in the water distribution system (node v_{10}) is also part of the electric power system, because it receives electric power from power line e_{10}. The water treatment facility in v_9 is physically connected with electric power node v_7. One could imagine a large site with an electric power supply at the location of the offices, but the power for the water treatment process is supplied by cables on-site, between the office and the water treatment process. Node v_5 represents an outlet in the house to which, for example, an appliance is connected. The last node is the electric vehicle charging station in node v_3. The electric vehicle charging station is naturally also part of the transportation system infrastructure.

The *transportation system* consists of four nodes and four edges. As mentioned, node v_3 is an electric charging station that facilitates electric vehicle charging. Nodes v_2 and v_4 are parking lots that facilitate the storage of vehicles. Lastly, Node v_1 is a private parking location at the house. Note that edge e_4 differs from the regular roads; it is a footpath. The transportation system thus contains two modes of transportation. As a result of the two modalities in the transportation system, the end users of the system are able to switch between these modalities. Consequently, the nodes v_1 and v_4 facilitate the transition from travelling on foot to travelling by EV and back.

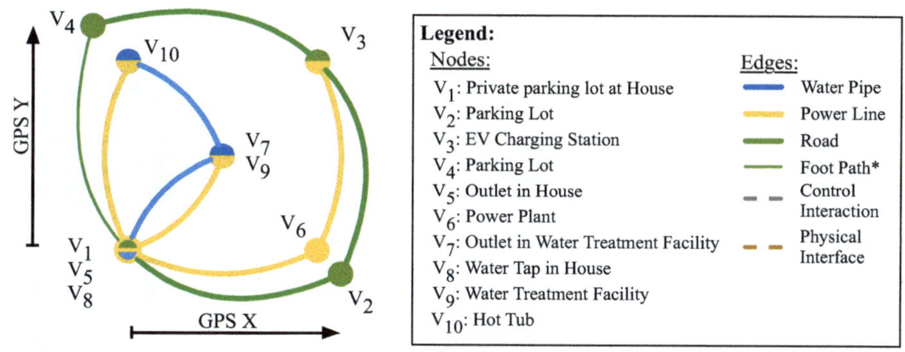

Fig. A.2 2D Presentation of a Hypothetical Four-Layer Network: It represents transportation, electric power, and water distribution infrastructure from Figure A.1

The *control structure* of the four-layer network includes an end user in v_{15} that controls the nodes related to the house. Additionally, node v_{11} is a separate end user who uses the hot tub. The water treatment facility is managed by the water utility, and similarly, the electric power utility manages the power plant. Note that the electric power utility works with the operator of the electric vehicle charging station to coordinate charging loads on the power grid. Finally, node v_{14} represents the operator of the two parking lots.

Figure A.2 provides a two-dimensional representation of Figure A.1, without the control structure. The figure superimposes nodes that have physical couplings, such as the house (Nodes v_1, v_5, and v_8), and the water treatment facility (Nodes v_7 and v_9). Other nodes, such as the hot tub (Node v_{10}) and parking lot 2 (Node v_4), are not superimposed to emphasize that they are distinct. This new representation emphasizes the spatial topology of the infrastructure network and lends itself for a clearer comparison later in the appendix.

A.1 System Concept

The first model in hetero-functional graph theory is the system concept. The system concept maps system function onto system form. This section first discusses the system form of the four-layer network and then continues the definition of its system function. Finally, the mapping of the processes onto the resources is performed with the knowledge base.

The resource architecture adheres to the LFES architecture as described in Chapter 4. Figure A.3 presents the resource architecture for the four-layer network. The system classifies 11 resources in nine interface classes. Note that parking lot 1 and parking lot 2 have different connections. In contrast to parking lot 1, parking lot 2 connects to a footpath and a road and should facilitate modality change of passengers between EV and traveling on foot. The set of resources in the four-layer network is: $R = M \cup B \cup H$, of size: $\sigma(R) = \sigma(M) + \sigma(B) + \sigma(H) = 6+1+10 = 17$.

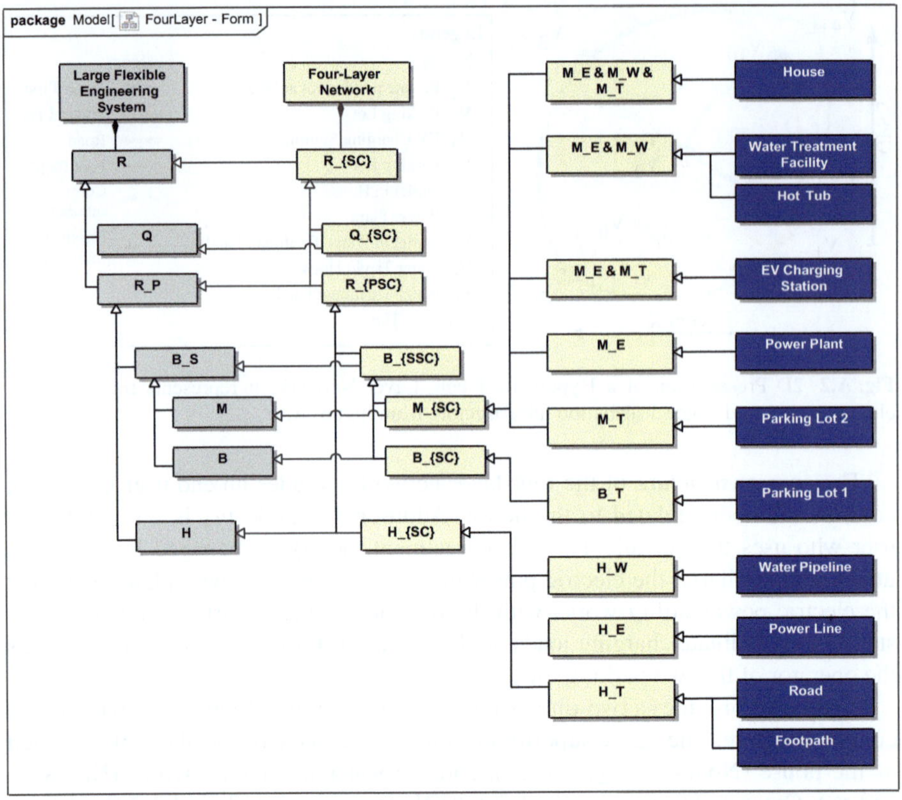

Fig. A.3 SysML Block Definition Diagram of the Four-Layer Network as a specialization of the LFES meta-architecture

The four-layer network activity diagram, as presented in Figure A.4, is similar to the activity diagram for the Trimetrica test case (Figure 5.11 on Page 112). The most notable difference is the addition of the modality "carry pedestrian." In order to facilitate sequences with a modality change, two transformation processes are added that allow for entering and exiting the electric vehicle. The set of transformation processes P_μ has size seven. The set is P_μ = {consume cold water, consume electric power, enter EV, exit EV, treat water, consume hot water, generate electric power}. The set of transportation processes P_η has size $\sigma(B_S)^2 = 49$. The set of holding processes P_γ has size six. It contains: P_γ = {potable water, electric power, park EV, charge EV, discharge EV, carry pedestrian}.

The capabilities are calculated as the mapping of system processes onto system resources. The knowledge base J_S is calculated using Equation 4.16 on Page 50. The size of J_S is 301×17, and it contains 34 capabilities. The SysML block diagram in Figure A.5 presents the unique set of capabilities for each of the resource classes in the four-layer network. The capabilities are visualized in Figure A.6, in which their locations correspond to the location of their associated resources.

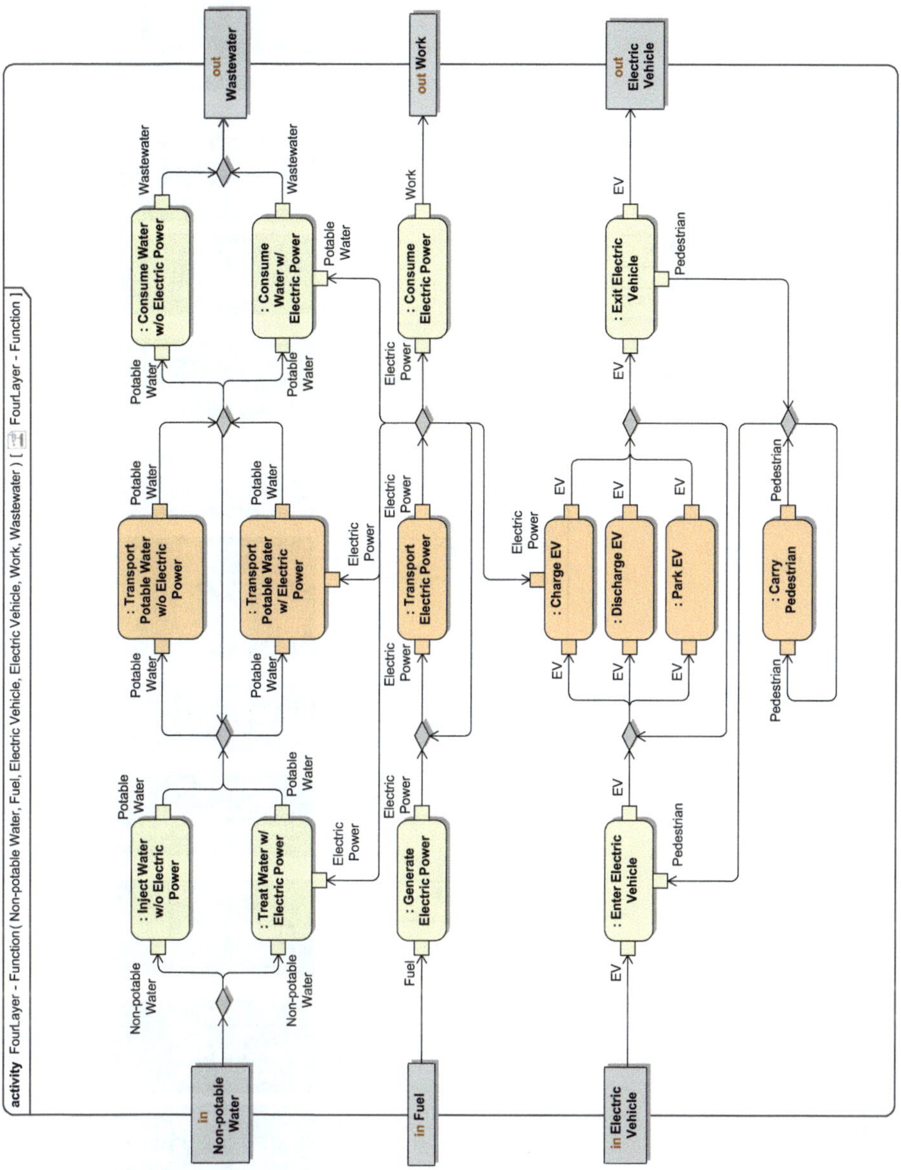

Fig. A.4 Activity Diagram of the Four-Layer Network Reference Architecture: The four operands are water, electric power, EV, and pedestrians

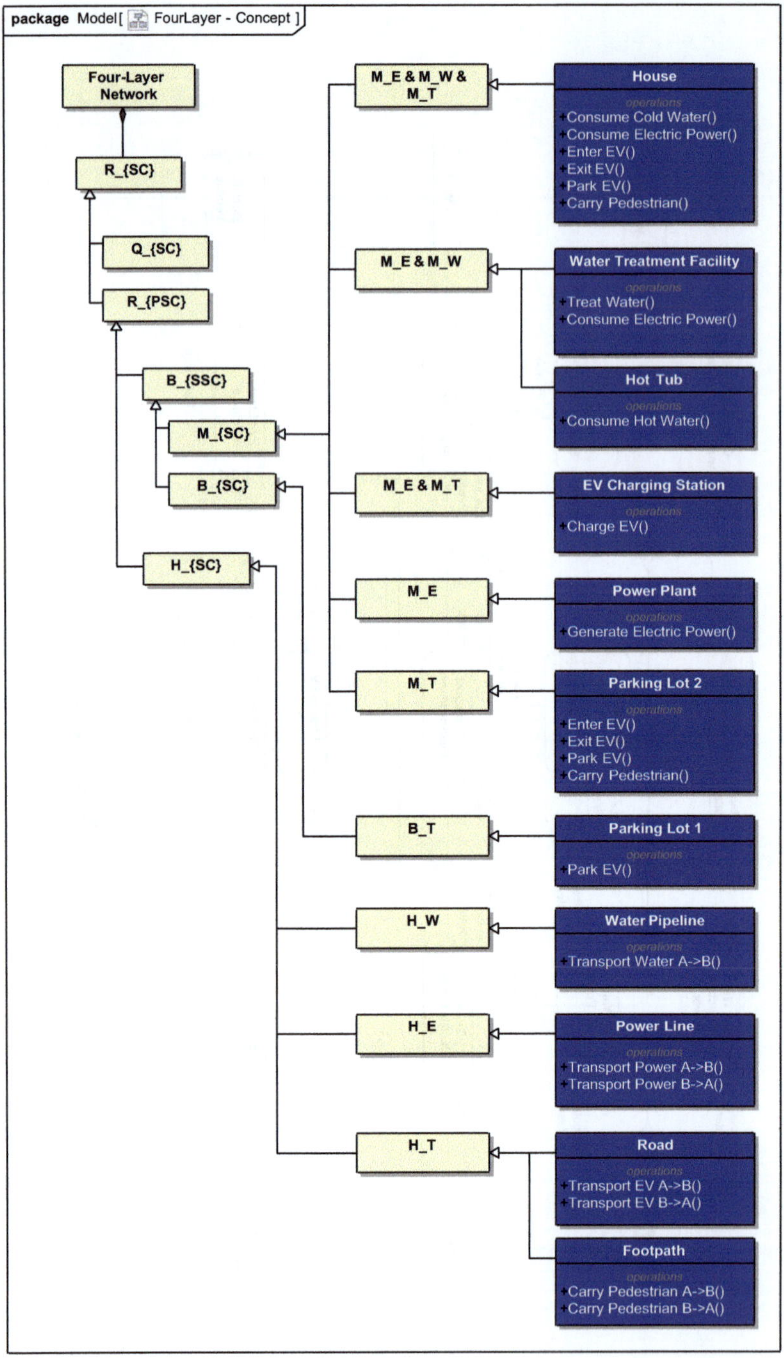

Fig. A.5 SysML Representation of the System Concept for the Four-Layer Network: This figure contains the unique set of capabilities for each of the resource classes

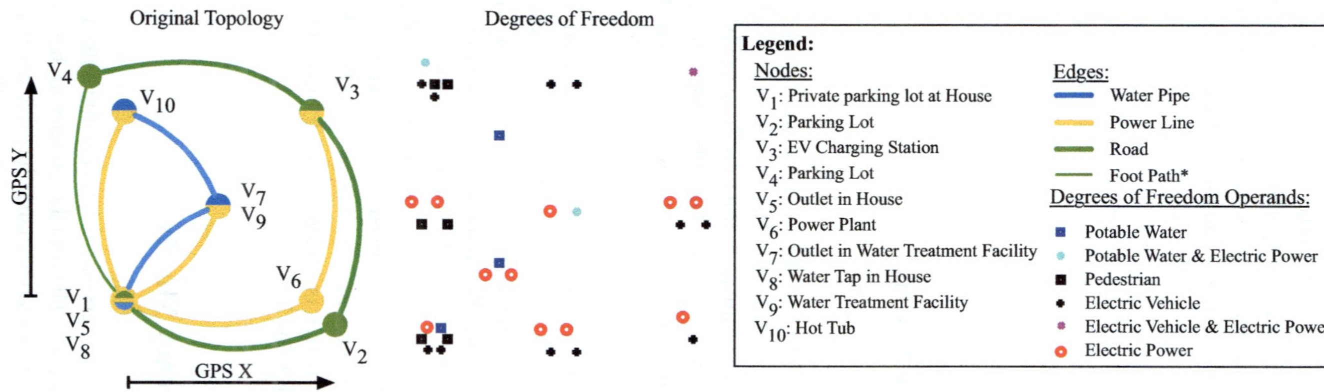

Fig. A.6 Topological presentation of the Degrees of Freedom in the Four-Layer Network: The original network topology is presented on the left, and the structural degrees of freedom are presented on the right. The degrees of freedom are classified by their operand type, e.g. "exit EV at house" has operand Electric Vehicle, whereas "enter EV at house" has operand Pedestrian

A.2 Hetero-functional Adjacency Matrix

The hetero-functional adjacency matrix couples the capabilities to represent the logical order of physical capabilities in a system. The hetero-functional adjacency matrix A_ρ and the system sequence degrees of freedom are calculated by Equations 4.28 and 4.29 on Page 58. For these equations, Table 4.3 on Page 57 presents the physical continuity constraints, and the design pattern in Figure A.4 represents the functional sequence constraints. The size of A_ρ is: $\sigma(R)\sigma(P) \times \sigma(R)\sigma(P) = 5117 \times 5117$. The projected matrix $\widetilde{A}_\rho$ has size $\sigma(\mathcal{E}_S) \times \sigma(\mathcal{E}_S) = 34 \times 34$. The number of sequence-dependent degrees of freedom is 79. Figure A.7 presents the visualization of the hetero-functional adjacency matrix superimposed on the degrees of freedom.

A.3 Controller Agency Matrix

The controller agency matrix serves to differentiate between two systems of equivalent capabilities but different control structure. Figure A.1 introduces six controller agents for the four-layer network. Each of these controller agents controls one or more nodes. However, the figure does not introduce a cyber-physical interface between the edges and the controller agents. The following control relationships are assumed: Node v_{12} controls the water pipelines, node v_{13} controls the power lines, and node v_{15} controls the footpath, and roads (it represents individuals that use the transportation network).

The independent controller agency matrix $\overline{A}_Q$ is defined in Definition 4.18 on Page 62. It maps the physical resources onto the cyber-resources Q. The set of cyber-resources has size six, and contains: $Q = \{$Person Operating Hot Tub, Water Treatment Facility Operators, Power Utility, Parking Lot Operator, Person in House, Operator of EV Charging Station$\}$. Consequently, the size of the independent controller agency matrix is $\sigma(R_P) \times \sigma(Q) = 17 \times 6$. The controller agency matrix is now mapped onto the capabilities, as demonstrated for Trimetrica in Equation 5.14 on Page 135. The size of the matrix now is: $\sigma(\mathcal{E}_S) \times \sigma(Q) = 34 \times 6$. The number of filled elements is 34. The matrix is visualized as a bipartite graph between the cyber-resources and the capabilities in Figure A.8.

A.4 Controller Adjacency Matrix

The controller adjacency matrix serves to describe the interactions between the cyber-resources as cyber-interfaces. The four-layer network has six cyber-resources, and consequently, its controller adjacency matrix has size 6×6. Figure A.1 contains a single cyber-interface, between nodes v_{13} and v_{16}. Since the figure does not

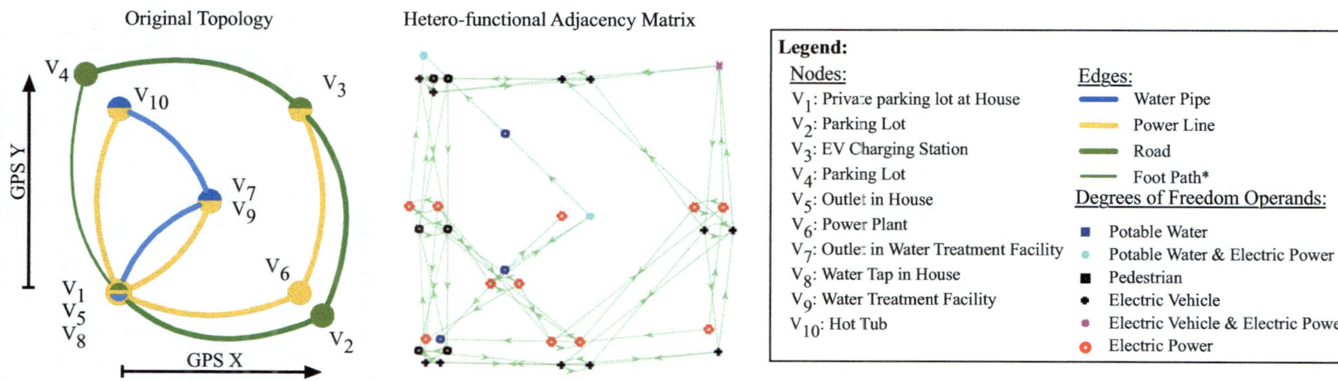

Fig. A.7 Topological presentation of the Hetero-functional Adjacency Matrix: The original network topology is presented on the left, and the structural degrees of freedom and the hetero-functional adjacency matrix are presented on the right

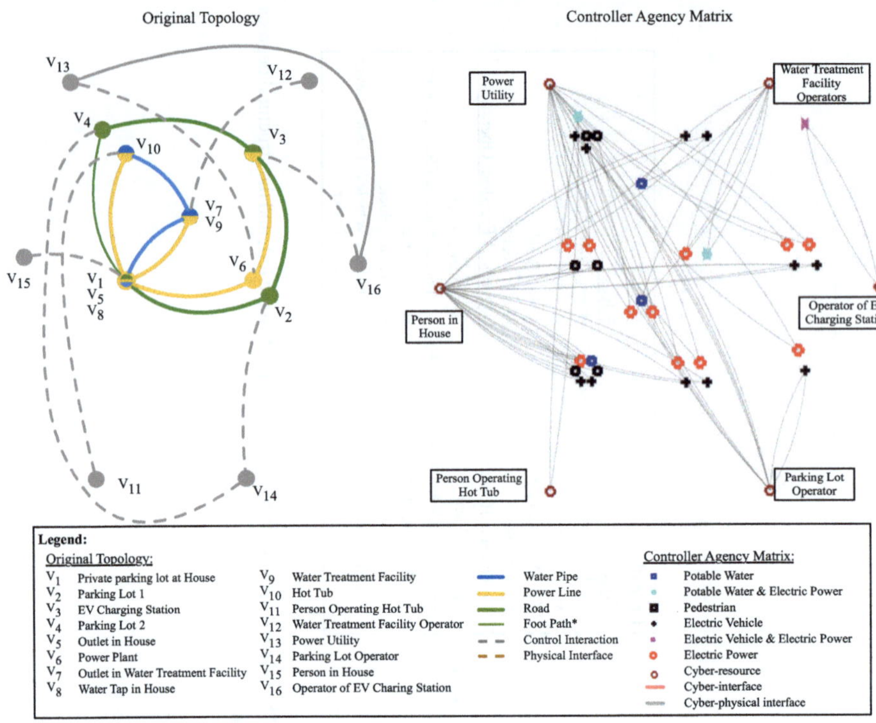

Fig. A.8 Controller Agency Matrix for the Four-Layer Network

specify the cyber-interface in greater detail, a bidirectional interface is assumed. The controller adjacency matrix, therefore, contains two filled elements. It is visualized in Figure A.9. The figure shows two interfaces, and four cyber-resources without any cyber-interfaces.

A.5 Service as Operand Behavior

The fifth model in hetero-functional graph theory describes the evolution of operands as they move through the physical infrastructure system. The system capabilities and their control structure have been described in the previous sections, and service model describes how the capabilities act on the operands.

The four-layer network performs three services, that are presented in Figure A.10 as service nets. The service *"deliver water"* consists of four transitions and one state. The state S_{1l_1} is "potable water." The set of service transitions is: $\mathcal{E}_{l_1} = \{$treat water(), maintain water(), consume hot water(), consume cold water()$\}$. The second service, *"deliver electric power,"* consists of six transitions and one state. The state S_{1l_2} is "electric power." The set of service transitions is: $\mathcal{E}_{l_2} = \{$generate electric

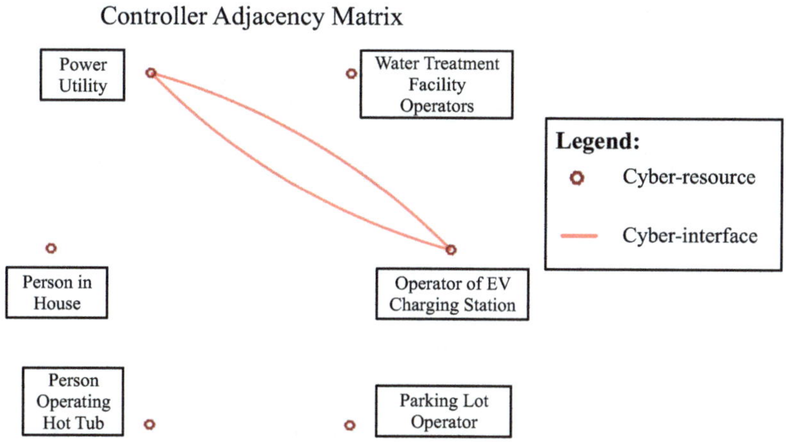

Fig. A.9 Controller Adjacency Matrix for the Four-Layer Network

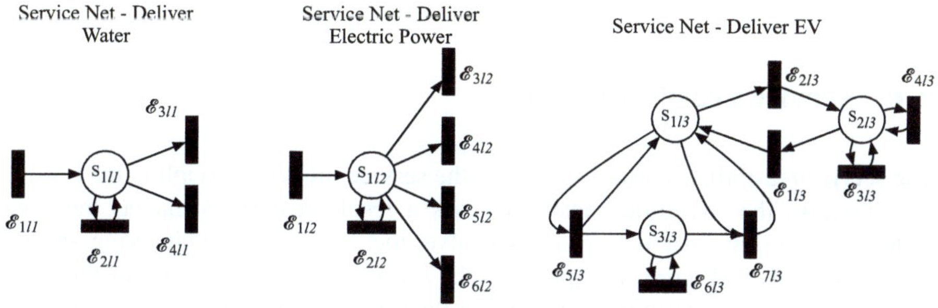

Fig. A.10 Service Nets for the Four-Layer Network: The four-layer network delivers three services: (1) deliver potable water, (2) deliver electric power, and (3) deliver EV

power(), maintain electric power(), treat water(), consume hot water(), consume electric power(), charge EV()}. The last service, *"deliver EV",* consists of three states and seven service transitions. Note that the name of this service does not cover all of its content, because the service also includes the movement of pedestrians. The set of the three states is: S_{l_3} = {pedestrian in EV, pedestrian outside EV, state-of-charge EV}. The set of service transitions is: $\mathcal{E}_{l_3}$ = {enter EV(), exit EV(), walk outside EV(), stay outside EV(), charge EV(), maintain EV(), discharge EV()}. Note that this service has a particularly interesting service net, as the pedestrian has state "in EV," while the EV operates. The EV cannot operate if the pedestrian is not in the EV.

Furthermore, the service nets are converted to service graphs, that show the adjacency of the service transitions. Figure A.11 shows the three service graphs for each of the services in the four-layer network. The service adjacency matrix for *deliver water* has a size of 4×4, with six filled elements that represent the adjacency.

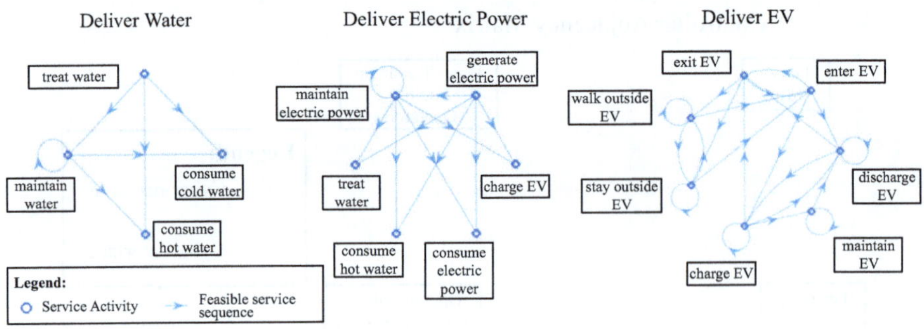

Fig. A.11 Service Graphs for the Four-Layer Network

The service adjacency matrix for *deliver electric power* has a size of 6 × 6, with ten filled elements. Finally, the service adjacency matrix for *deliver EV* has a size of 7 × 7, with 22 filled elements.[1]

A.6 Service Feasibility Matrix

The service feasibility matrices complete the service model by coupling the service transitions to the structural capabilities. As a result, it ensures that as the state of the engineering system's operands evolve, the state of the engineering system itself also evolves. Figure A.12 presents the service feasibility matrix for the four-layer network as a bipartite graph between the service transitions and the structural degrees of freedom. The service feasibility matrix is calculated for each of the operands by Equation 4.44. The service feasibility matrix for the service *deliver water* is Λ_1, which has a size of 4 × 301. When projected, $\widetilde{\Lambda}_1$ has a size of 4 × 34, with five filled elements. The service feasibility matrix for the service *deliver electric power* is Λ_2, which has a size of 6 × 301. When projected, $\widetilde{\Lambda}_2$ has a size of 6 × 34, with 14 filled elements. The service feasibility matrix for the service *deliver electric vehicle* is Λ_3, which has a size of 7 × 301. When projected, $\widetilde{\Lambda}_3$ has a size of 7 × 34, with 18 filled elements.

A.7 System Adjacency Matrix

The system adjacency matrix is the final model of hetero-functional graph theory. It integrates the six previous models to provide a holistic representation of the cyber-physical engineering system. Equation 4.67 on Page 82 provides the calculation of the projected system adjacency matrix $\widetilde{\mathbb{A}}$:

[1]Note that the service net contains two paths that connect $\mathcal{E}_{4l_3}$ and $\mathcal{E}_{6l_3}$.

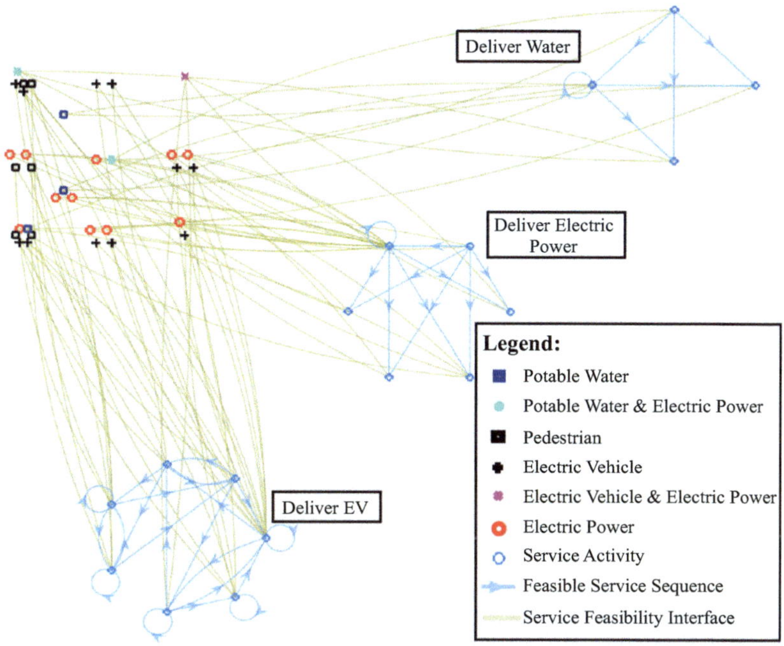

Fig. A.12 Service Feasibility Matrix as a Bipartite Graph for the Four-Layer Network

$$\widetilde{\mathbb{A}} = \begin{bmatrix} \mathbb{A}_L & \widetilde{\mathbb{A}}_{L\rho} & \mathbf{0} \\ \widetilde{\mathbb{A}}_{\rho L} & \widetilde{A}_\rho & \widetilde{\mathbb{A}}_{\rho C} \\ \mathbf{0} & \widetilde{\mathbb{A}}_{C\rho} & A_C \end{bmatrix} \tag{A.1}$$

The matrix $\widetilde{A}_\rho$ is the hetero-functional adjacency matrix, as calculated in Section A.2. Matrix A_C is the controller adjacency matrix, as calculated in Section A.4. Matrix $\mathbb{A}_L$ is the concatenation of the service transition adjacency matrices. The matrices are calculated in Section A.5, and concatenated as demonstrated in Equation 4.68 on Page 84. Matrix $\widetilde{\mathbb{A}}_{\rho C}$ maps the independent cyber-resources to the structural degrees of freedom, as calculated in Section A.3. Finally, matrix $\widetilde{\mathbb{A}}_{L\rho}$ is the service feasibility matrix that maps the service transitions to the structural degrees of freedom, as calculated in Section A.6. Finally, Figure A.13 presents the System Adjacency Matrix for the four-layer network.

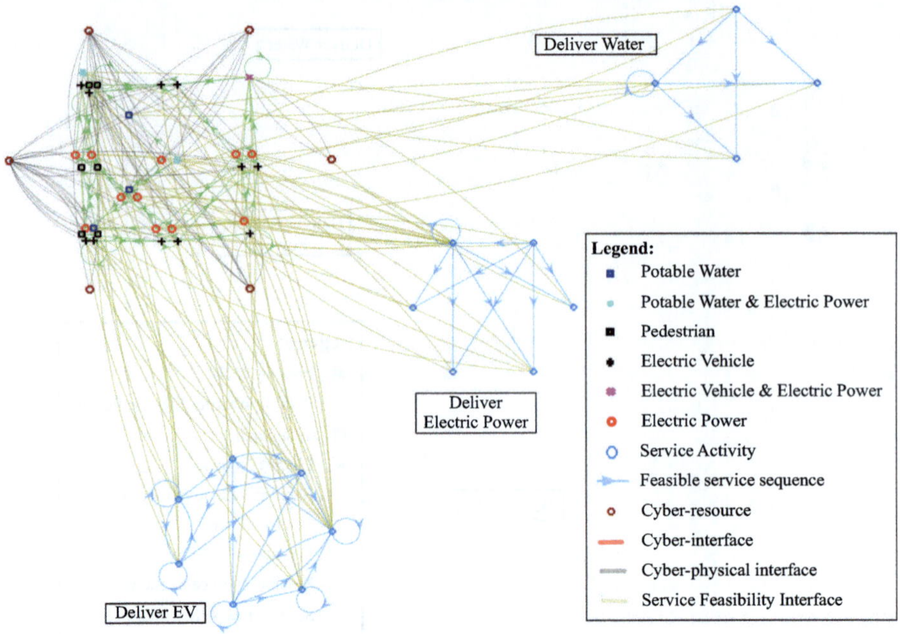

Fig. A.13 System Adjacency Matrix for the Four-Layer Network

List of Definitions

W. C. H. Schoonenberg et al., *A Hetero-functional Graph Theory for Modeling Interdependent Smart City Infrastructure*,
https://doi.org/10.1007/978-3-319-99301-0

Index

W. C. H. Schoonenberg et al., *A Hetero-functional Graph Theory for Modeling Interdependent Smart City Infrastructure*,
https://doi.org/10.1007/978-3-319-99301-0

MIX
Papier aus verantwortungsvollen Quellen
Paper from responsible sources
FSC® C105338

If you have any concerns about our products,
you can contact us on
ProductSafety@springernature.com

In case Publisher is established outside the EU,
the EU authorized representative is:
Springer Nature Customer Service Center GmbH
Europaplatz 3, 69115 Heidelberg, Germany

Printed by Libri Plureos GmbH
in Hamburg, Germany